THE LUCENT LIBRARY OF **SCIENCE AND TECHNOLOGY**

Plate Tectonics

by Kay Jackson

LUCENT BOOKS

An imprint of Thomson Gale, a part of The Thomson Corporation

THOMSON

GALE

Detroit • New York • San Francisco • San Diego • New Haven, Conn. • Waterville, Maine • London • Munich

For Doug

LIBRARY OF CONGRESS CATALOGING-IN-PUBLICATION DATA

Jackson, Kay, 1959–
    Plate tectonics / by Kay Jackson.
        p. cm. — (Lucent library of science and technology)
Includes bibliographical references and index.
ISBN 1-59018-105-0 (hardcover : alk. paper)
1. Plate tectonics—Juvenile literature. I. Title. II. Series.
QE511.4.J33 2005
551.1'36—dc22
                                                                        2004021916

Printed in the United States of America

Table of Contents

Foreword

"The world has changed far more in the past 100 years than in any other century in history. The reason is not political or economic, but technological—technologies that flowed directly from advances in basic science."

— Stephen Hawking, "A Brief History of Relativity," *Time*, 2000

The twentieth-century scientific and technological revolution that British physicist Stephen Hawking describes in the above quote has transformed virtually every aspect of human life at an unprecedented pace. Inventions unimaginable a century ago have not only become commonplace but are now considered necessities of daily life. As science historian James Burke writes, "We live surrounded by objects and systems that we take for granted, but which profoundly affect the way we behave, think, work, play, and in general conduct our lives."

For example, in just one hundred years, transportation systems have dramatically changed. In 1900 the first gasoline-powered motorcar had just been introduced, and only 144 miles of U.S. roads were hard-surfaced. Horse-drawn trolleys still filled the streets of American cities. The airplane had yet to be invented. Today 217 million vehicles speed along 4 million miles of U.S. roads. Humans have flown to the moon and commercial aircraft are capable of transporting passengers across the Atlantic Ocean in less than three hours.

The transformation of communications has been just as dramatic. In 1900 most Americans lived and worked on farms without electricity or mail delivery. Few people had ever heard a radio or spoken on a telephone. A hundred years later, 98 percent of American

4

homes have telephones and televisions and more than 50 percent have personal computers. Some families even have more than one television and computer, and cell phones are now commonplace, even among the young. Data beamed from communication satellites routinely predict global weather conditions, and fiber-optic cable, e-mail, and the Internet have made worldwide telecommunication instantaneous.

Perhaps the most striking measure of scientific and technological change can be seen in medicine and public health. At the beginning of the twentieth century, the average American life span was forty-seven years. By the end of the century the average life span was approaching eighty years, thanks to advances in medicine including the development of vaccines and antibiotics, the discovery of powerful diagnostic tools such as X rays, the lifesaving technology of cardiac and neonatal care, improvements in nutrition, and the control of infectious disease.

Rapid change is likely to continue throughout the twenty-first century as science reveals more about physical and biological processes such as global warming, viral replication, and electrical conductivity, and as people apply that new knowledge to personal decisions and government policy. Already, for example, an international treaty calls for immediate reductions in industrial and automobile emissions in response to studies that show a potentially dangerous rise in global temperatures is caused by human activity. Taking an active role in determining the direction of future changes depends on education; people must understand the possible uses of scientific research and the effects of the technology that surrounds them.

The Lucent Books Library of Science and Technology profiles key innovations and discoveries that have transformed the modern world. Each title strives to make a complex scientific discovery, technology, or phenomenon understandable and relevant to the reader. Because scientific discovery is rarely straightforward, each title explains the dead ends, fortunate accidents, and basic scientific methods by which the research into the subject proceeded. And every book examines the practical applications of an invention, branch of science, or scientific principle in industry, public health, and personal life, as well as potential future uses and effects based on ongoing research. Fully documented quotations, annotated bibliographies that include both print and electronic sources, glossaries, indexes, and technical illustrations are among the supplemental features designed to point researchers to further exploration of the subject.

Introduction

Modern Tragedies and Ancient Forces

On December 22, 2003, an earthquake struck central California. The magnitude 6.5 quake originated near San Simeon, California. Only two people were killed and forty were injured. Ross Stein, a geologist with the U.S. Geological Survey, describes how the effects of the quake could be felt all over the globe: "For an earthquake this size, every single sand grain on the planet dances to the music of those seismic waves. You may not be able to feel them, but the entire planet is rung like a bell."[1] Just four days later a magnitude 6.5 quake also hit the historic city of Bam, Iran.

Many of Bam's eighty thousand citizens lived in homes built mostly of adobe, or mud brick. In fact, Bam's famous Old City is the site of the world's oldest and largest mud brick structure, a fortress complex of towers, stables, and domes. After the tremors stopped, 85 percent of the city lay in jumbled ruins. The two-thousand-year-old citadel crumbled into a dusty heap, and entire city blocks were destroyed. The human toll was even more staggering. Throughout the following weeks, the count of the quake's victims rose to 43,200 dead, 30,000 injured, and 75,000 homeless.

Family members sort through the ruins of their home after a magnitude 6.5 earthquake struck the city of Bam in Iran in December 2003.

A glimpse of the disaster's emotional toll is found in the story of Mohammed Karimi of Bam, who lost his wife and four-year-old daughter in the quake. Tearfully, he recalls the last conversation with the little girl. "Last night before she went to sleep she made me a drawing and kissed me four times. When I asked, 'Why four kisses?' she said, 'Maybe I won't see you again, Papa.'"[2]

The threat to Bam is a common one, shared by other communities that are built along fault lines, including the cities along the San Andreas Fault in California. Though on opposite sides of the planet, both earthquakes were the result of plate tectonics, the slow yet relentless movement of the huge pieces of Earth's crust.

The theory of plate tectonics explains why and how crustal plates, or immense rock slabs, move around on Earth. "Tectonic" comes from the ancient Greek word *tekton*, which means "builder"; and over billions of years, tectonic forces have been at work producing the features of the planet: tall mountains, volcanoes, broad plateaus, deep canyons, and even the cold ocean floor. As the crustal plates have wandered very, very slowly over the surface of Earth, climates have changed and life forms have evolved to adapt to new conditions.

As the California and Iran quakes show, plate tectonics can drastically affect people's lives. People, however, cannot influence such forces. Earthquakes, volcanoes, and tsunamis are violent and sometimes deadly natural disasters caused by plate tectonics. Humans can try to manage the effects of such disasters, but they cannot stop an earthquake or change a tsunami's course. Yet, for all its potential for destruction, plate tectonics is responsible for the life in all its varieties found on the planet.

Oceanographer and marine geologist Ellen Prager sums up what Earth would be without plate tectonics: "The third planet from the sun would be a very

Spectators watch a plume of steam and ash rise from Mount Saint Helens in Washington. The volcano's 1980 eruption killed fifty-seven people.

dull place indeed."[3] Despite the destruction associated with earthquakes, scientists like Tim Rockwell, a professor of geological sciences at San Diego State University in California, believe that earthquakes are necessary to the planet. "Without earthquakes, without motion on faults, we wouldn't have mountains, we wouldn't have hills. What we would have, over time, is a lot of erosion that would reduce all of this material to sand. So earthquakes and plate tectonics are a vibrant and critical element that keeps this Earth alive. They're part of the pulse. They're part of the breathing of the planet that makes it a great place to live."[4]

With a wide array of technologies, earth scientists explore the planet and search for clues to the mysteries of Earth's evolution. Cameras provide images from the sunless depths of the deep ocean trenches. Satellite-mounted radar equipment produces three-dimensional maps of shifting land. Small hand-held global positioning system units receive data from a constellation of satellites to pinpoint the movement of the world's tectonic plates. Powerful computers generate models that help scientists to re-create Earth's prehistoric past and predict its distant future.

Chapter 1

Continental Drift and Plate Tectonics

Before 1960 geologists were certain that continents did not move. They believed that continents were permanent features rooted deep in Earth's crust. By 1970, however, most agreed that the continents were not only moving but that they had been colliding and then drifting apart for millions of years. The theory of plate tectonics helps to explain how and why continents move.

Plate Tectonics

The theory of plate tectonics is crucial to understanding Earth's physical systems. As important as the theory of evolution is to the life sciences, plate tectonics sheds light on how the planet and its features were formed. To earth scientists like W. Kenneth Hamblin, a geologist and professor at Brigham Young University in Utah, it "explains the many aspects of the interrelationships of volcanoes, earthquakes, climate change and even the evolution of life itself. In brief, it provides a single unifying theory of Earth's dynamics. Essentially everything about our planet is related either directly or indirectly to plate tectonics."[5] It is, however, more than a single, isolated theory.

Plate tectonics brings scientists together from diverse fields. For example, geologists work alongside paleontologists to create a picture of ancient life in prehistoric climates. Seismologists collaborate with geophysicists to develop a better understanding of earthquakes. Plate tectonics has allowed scientists to connect ideas and information that once seemed unrelated. Tanya Atwater, a professor of tectonics at the University of California at Santa Barbara, found that "the best part of the plate business is that it has made us all start communicating. People who squeeze rocks and people who identify deep ocean nannofossils and people who map faults in Montana suddenly all care about each other's work."[6] Like links in a chain, pieces of information from each specialty contribute to a more comprehensive knowledge of the planet.

In 1915 meteorologist Alfred Wegener proposed the concept of continental drift, which became the basis of the theory of plate tectonics.

While many scientists helped to develop the theory of plate tectonics, Alfred Wegener is credited with laying its foundation.

Alfred Wegener and Drifting Continents

From Leonardo da Vinci in the fifteenth century to the world explorer Alexander von Humboldt in the eighteenth century, people looked at maps and noticed that the west coast of Africa seemed to match the east coast of South America. This mystery fascinated Alfred Wegener, a young German meteorologist. He wanted to know why the two continents appeared to fit like the pieces of a jigsaw puzzle. In a letter to his fiancé, Wegener wrote, "Doesn't the east coast of South America fit exactly against the west coat of Africa as if they had once been joined? . . . This is an idea I'll have to pursue."[7] His idea became the theory of continental drift, a notion so revolutionary that Seiya Uyeda, a Japanese geophysicist, describes it as "a spark that generated a new view of the earth."[8]

Wegener first proposed the concept of mobile continents in a book published in 1915. Its title in English is *The Origins of Continents and Oceans.* According to Wegener, the continents had been moving across the face of the earth for hundreds of millions of years. Sometimes continents collided, and mountains were formed. At other times continents split apart, and seas filled in the divides. This rifting process explains the corresponding shapes of the coastlines of South America and Africa. Throughout the book Wegener carefully built his case with evidence from multiple branches of science. Paleontology, the study of prehistoric life by examining fossils, proved to be especially valuable in providing confirmation that the continents moved.

Citing observations going back to von Humboldt, Wegener pointed out that rocks taken from the eastern shore of South America were identical to those found on the western coast of Africa. Samples from both places also contained fossils of the same plants and animals.

Not only did the shapes of Africa and South America match, so did their fossils and rocks. Wegener felt this proved that the continents were once connected. He imagined the matching rocks and fossils as printed lines running across a piece of paper and wrote, "It is just as if we were to refit the torn pieces of a newspaper by matching their edges and then check whether the lines of print run smoothly across. If they do, there is nothing left but to conclude that the pieces were in fact joined in this way."[9] For Wegener, there was no other logical explanation. The fossil record proved to support continental drift in another way.

Wegener used his knowledge and experiences as a meteorologist to connect prehistoric climates and fossil records with drifting continents. Fossils of tropical plants like the fern *Glossopteris* were found in frozen Antarctica while deposits left by glaciers had been found in the warm, tropical regions of southern Africa, India, and South America. Obviously, the past climates of continents differed from their present ones. This is because the continents have moved: Antarctica once had been closer to the equator, while Africa, India, and South America had been located nearer to the South Pole.

Wegener concluded the continents had been united into one enormous landmass until about 250 million years ago. He called this supercontinent Pangaea. About 200 million years ago Pangaea slowly broke up into two large continents: Gondwanaland in the Southern Hemisphere, which would come to be South America, Africa, peninsular India, Australia, and Antarctica; and Laurasia north of the equator, which consisted of the land that would become North America, Europe, and Asia. Throughout millions of years Gondwanaland and Laurasia slowly divided into smaller bodies. At times the continents plowed into each other and portions of landmasses combined; on one such occasion India crashed into Asia, to which it remains joined. In another major geological event, Gondwanaland split into two distinct landmasses, South America and Africa.

Rejection and an Icy Death

For all the logic of its main conclusions, *The Origins of Continents and Oceans* did not offer a realistic ex-

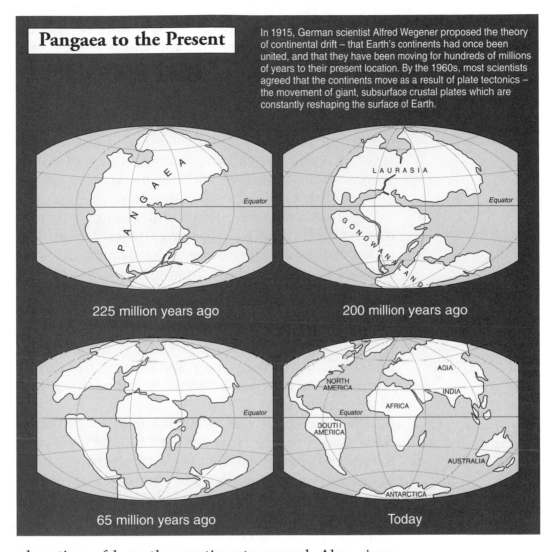

Pangaea to the Present

In 1915, German scientist Alfred Wegener proposed the theory of continental drift – that Earth's continents had once been united, and that they have been moving for hundreds of millions of years to their present location. By the 1960s, most scientists agreed that the continents move as a result of plate tectonics – the movement of giant, subsurface crustal plates which are constantly reshaping the surface of Earth.

225 million years ago

200 million years ago

65 million years ago

Today

planation of how the continents moved. Also, since Wegener was a meteorologist, most geologists saw him as an outsider without the proper training or qualifications to propose such a radical notion. Further, in order to accept continental drift geologists would have to forsake their long-held notions about the earth. Rollin Chamberlain, a well-respected geologist from the University of Chicago, claimed in 1926, "If we are to believe Wegener's hypothesis, we must forget everything that has been learned in the past seventy years and start over again."[10] Most scientists were not willing to give up the idea of fixed continents for moving ones, and continental drift failed to gain acceptance during Wegener's lifetime.

Alfred Wegener proved to be a stubborn man. Despite rejection and ridicule from the scientific community, he continued to pursue evidence for his theory and revise *The Origins of Continents and Oceans.* Wegener also continued with his meteorological career until his death in 1930 at age fifty.

Wegener died in Greenland while leading a meteorological team that was studying the jet stream. The expedition had a permanent base camp on the west coast of the island while a temporary research station was located deep in the interior. Wegener feared that the research station's two-man crew did not have enough supplies to last through the winter. However, since the station lacked a radio, he could not confirm nor deny his fears. Despite the uncertainties, Wegener decided to risk the dangerous trek to the interior station.

In November, he and two other men set out from the coastal base camp to make the 250-mile journey to the research station. During the difficult march, most of the food they were transporting was lost. When Wegener's team reached the research station, they realized that there was not enough food for the five of them to live through the winter. There were, however, enough supplies for three people. This meant that two men would have to leave the research station and return to the base camp if anyone was to survive.

With few supplies and no dog sleds, Wegener and his Inuit guide elected to attempt the grueling trek. Neither man survived. Expedition members found Wegener's body the next spring neatly bundled in his sleeping bag. Those who found him remarked that "Wegener had a peaceful, almost smiling, look on his face—the look of a man who died knowing he had done what was right."[11]

Drift and Plate Tectonics

Alfred Wegener's legacy endures in the theory of plate tectonics. He brought together ideas, theories, and information from many different branches of earth science to form the theory of continental drift, the precursor to plate tectonics. For Wegener, this approach was fundamental to understanding Earth. "It is only by combining the information furnished by all the earth sciences that we can hope to determine truth here, that is to say, to find the picture that sets out all the known facts in the best arrangement and that therefore has the highest degree of probability."[12]

New discoveries did change the theory of continental drift as originally proposed by Wegener. This time the evidence did not come from rocks and fossils found on land. Instead, the ocean floor provided the answer to the riddle of drifting continents. Though humans have called different areas of Earth's ocean by many names, in reality, there is only one ocean. In the early 1950s geologists aboard research ships outfitted with laboratories and sophisticated instruments began a series of historic voyages to probe the depths of the world's ocean, and a complete picture of the ocean floor gradually emerged.

Mapping the Ocean Floor

In the years following World War II, Maurice Ewing, a geologist from Columbia University, devised a technique that utilized sound waves to determine the thickness of sediment layers that cover the seafloor. Ewing's team would put out to sea in a research ship, locate the area to be investigated, and begin detonating TNT to create sound waves. This was exciting and potentially dangerous work. One of Ewing's research crew, paleontologist David Ericson, describes the experience:

> I remember one used to hold the fuse in one's teeth, sitting on the afterdeck. The charge—a

This 1948 photo shows Maurice Ewing aboard his research vessel. Ewing's seismic surveys led to the discovery of the mid-ocean ridge, a vast underwater mountain chain.

half pound of TNT—was in one hand and a lighter in the other. One wasn't supposed to put the fuse in until the last minute. Then you flung it. This continued at half-hour intervals day and night.

[Ewing] would seize every opportunity to get information. There would be two-day intervals where he was going continuously. If the topography became surprising the ship would circle and try to outline it. With him you were either going to sink the ship with too much explosive or discover something interesting.[13]

Later, the TNT technique was refined so that charges of TNT were attached to a long cable that was strung out behind the ship. On command from

Ewing, a switch detonated the TNT in a regular sequence. The explosions produced sound waves that echoed off of the ocean floor. A waterproof sound recorder also trailed behind the ship and it inscribed on rolls of paper the patterns produced by the sound waves. When the experiment was complete, the crew would wind in the recorder and retrieve the paper tape. Recorded on the roll was a chart showing the different layers of rock of the ocean floor. Along with more conventional techniques to measure the depth of topographical features, Ewing's seismic surveys revealed an amazing feature hidden below the waves—Earth's greatest mountain system, the mid-ocean ridge.

The Mid-Ocean Ridge

The mid-ocean ridge is an enormous underwater mountain chain that wraps around Earth like the seam of a baseball. The ridge stretches across the face of the globe for more than thirty-eight thousand miles. Its peaks have an average height of fifteen thousand feet above the ocean floor with some mountains rising more than thirty thousand feet. More than revealing the existence of the planet's most prominent landform, the discovery of the mid-ocean ridge provided a key link in the formation of the theory of plate tectonics.

While studying the Atlantic section of the mid-ocean ridge, Marie Tharp, an oceanographic cartographer, uncovered a remarkable feature—a rift valley that ran through the center of the ridge for its full length. On either side of the valley lay similar systems of mountains and foothills. Tharp next imposed a map of earthquake activity over the map of the rift valleys. She and others were startled to see that the centers of the earthquakes fell on or near the valley.

The rift valley of the mid-ocean ridge provided an important clue to Harry Hess, a geophysicist who believed that continents moved. He theorized that rift valleys mark the spot where Earth's crust is pulling apart, or spreading. His idea of seafloor spreading, a new wrinkle in the continental drift theory, added another crucial piece to the plate tectonics puzzle.

Commander Hess

During World War II Hess had commanded a transport ship that moved marines and supplies throughout the South Pacific. His duties included using the ship's sonar equipment both to avoid enemy submarines and to contribute to the mapping of the ocean floor. Sonar, or SOund NAvigation Ranging technology, allows researchers not only to bounce

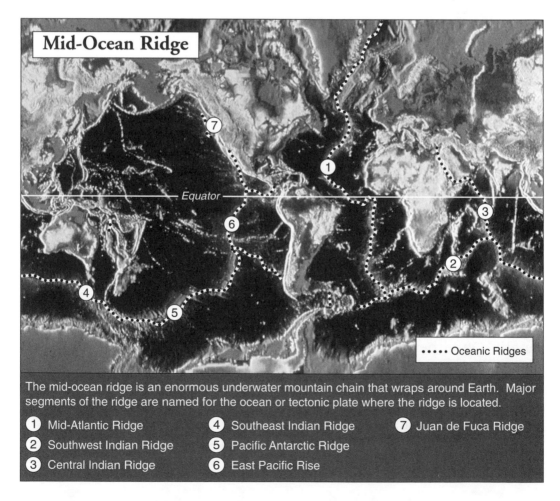

Mid-Ocean Ridge

Equator

····· Oceanic Ridges

The mid-ocean ridge is an enormous underwater mountain chain that wraps around Earth. Major segments of the ridge are named for the ocean or tectonic plate where the ridge is located.

1 Mid-Atlantic Ridge   4 Southeast Indian Ridge   7 Juan de Fuca Ridge
2 Southwest Indian Ridge   5 Pacific Antarctic Ridge
3 Central Indian Ridge   6 East Pacific Rise

sound waves off underwater objects but also records the time the waves take to travel back to the ship. From this information it is possible to calculate the distance from the object to the ship.

In a letter to a fellow geologist and friend, Hess wrote about his scientific successes while aboard the transport. "Have been able to get about a dozen or more traverses [survey lines] across deeps and can outline their course pretty well from Iwo [Jima] to Palau. Have four across the Mindanao deep, too. We filled in a lot of blank spots on the charts."[14]

Extinct Underwater Volcanoes Point to New Theory

As his ship proceeded from island to island, Hess mapped numerous flat-topped mountains that he named guyots after Arnold H. Guyot, an American geographer and geologist. The guyots were remnants of extinct volcanoes that had once risen above the ocean's surface but were now miles beneath it. Hess reasoned that wind and water had gradually worn down the sharp peaks, but he could not yet explain why the flat tops of the mountains came to be found so far below the surface of the ocean. He would have to wait for the end of the war to find that answer.

After World War II Hess returned to the Pacific once again. This time he commanded a research vessel equipped to explore the bottom of the ocean. Hess located the mysterious guyots and collected samples of rocks and sediments from their tops and sides. He then analyzed the samples to determine their ages. Hess had predicted that the guyots would be 500 million years old. He was surprised to find that they were only 180 million years old or even younger. When Hess realized his previous theories did not match the new data, he began to construct a new view of the ocean floor. By combining information about the mid-ocean systems of ridges and rifts with his own exploration of the seafloor, Hess constructed a theory to account for his surprising findings.

In 1960 Hess proposed that the ocean's floor was continually being created and then recycled by a process he called seafloor spreading. The rift valleys of the different sections of the mid-ocean ridge mark spots where Earth's surface is thin and weak. Hot molten rock from deep in the earth squeezes up through these weak areas. The molten rock, called magma, spreads away from the ridges and cools to form a new layer of ocean floor. Eventually, the old seafloor reaches an ocean trench and is pulled down deep below the surface by gravity. Hess believed that it took about 180 million years for rocks to travel from the mid-ocean ridge to the deep sea trenches. His theory of seafloor spreading, therefore, explains both the age of the seafloor and the thinness of its sediments. Hess also inspired other scientists to keep searching for answers below the waves. While studying the rocks found on the ocean floor, geologists made another surprising discovery.

Tiny Compasses Produce a Zebra Stripe

During the 1920s Japanese geologist Motonari Matuyama proved that Earth's magnetic poles had periodically switched direction throughout geological history. From measuring very small magnetic readings in rocks, scientists had learned that rocks have microscopic crystals that act like tiny compasses. These crystals form as the molten rocks deep in the earth begin to rise to the surface and cool. The minute crystals align to Earth's positive magnetic pole. In this way the magnetic readings of Earth are recorded in the rocks. Today, compasses point north, and if a rock's polarity aligns to the north, it has a normal polarity. However, if a rock's polarity is oriented to the south, it has a reversed polarity. While studying ancient volcanic rocks, Matuyama discovered that some rocks did not align to the north; in-

stead, their tiny crystalline compasses pointed south. Matuyama proved that the poles had periodically switched direction throughout geological history. Since then geologists found that there have been 171 magnetic reversals over the past 75 million years.

In the 1950s marine geologists observed patterns of magnetic reversals in the rocks on the seafloor. In the decades following World War II magnetometers were used to measure the direction and strength of a rock's magnetism. As the research ships towed the magnetometers through the water, the magnetic readings of the rocks on the seafloor were recorded on long strips of paper. A pattern began to emerge on the paper strips; white bands of normal polarity alternated with black bands of reversals. These sequences formed a "zebra stripe" pattern that intrigued several geologists. They believed the zebra stripes were evidence that the seafloor moved.

Fred Vine and Drummond Matthews at the University of Cambridge and Lawrence W. Morley of the Canadian Geological Survey proposed in the 1960s that the zebra stripes were evidence of magnetic

This computer model generated from sonar waves bounced off the seafloor reveals the geomagnetic orientation of a section of the mid-ocean ridge.

reversals that occurred as the seafloor spread out from the mid-ocean ridges. They saw the seafloor as a giant tape recorder that preserved Earth's switches in polarity. When samples were taken from both sides of a ridge system, the magnetic patterns mirrored each other. Vine, Matthews, and Morley also found the striped patterns at mid-ocean ridges all over the globe.

The Perfect Profile

In 1965 the research ship *Eltanin* took samples from the seafloor on both sides of a ridge in the eastern Pacific. Physicist Jim Heirtzler and his team from the Lamont Geological Observatory, a part of Columbia University in New York, analyzed the magnetic data drawn from the rock samples and created a profile, or diagram, for each. One profile in particular, a sample called *Eltanin*-19, caught everyone's attention.

The magnetic recordings astounded earth scientists. Tanya Atwater was beginning her long career in geology and remembers her amazement when she first saw the pattern created by *Eltanin*-19. "In his talk [Jim Heirtzler] put up the *Eltanin*-19 magnetic anomaly profile—still, to this day, the clearest, most beautiful, and symmetrical profile in the world—and made the case for seafloor spreading. It was as if a bolt of lightning had struck me. My hair stood on end."[15] Other scientists were struck by the near perfect symmetry and its implications.

Atwater and others realized that if the seafloor was moving, then so were the continents. Nearly fifty years after Alfred Wegener first proposed his theory of continental drift, scientists finally began to accept that continents moved. Wegener had been right.

Deep Sea Drilling

In the late 1960s rock samples taken from the seafloor confirmed the connection between seafloor spreading and the zebra stripe patterns of magnetic

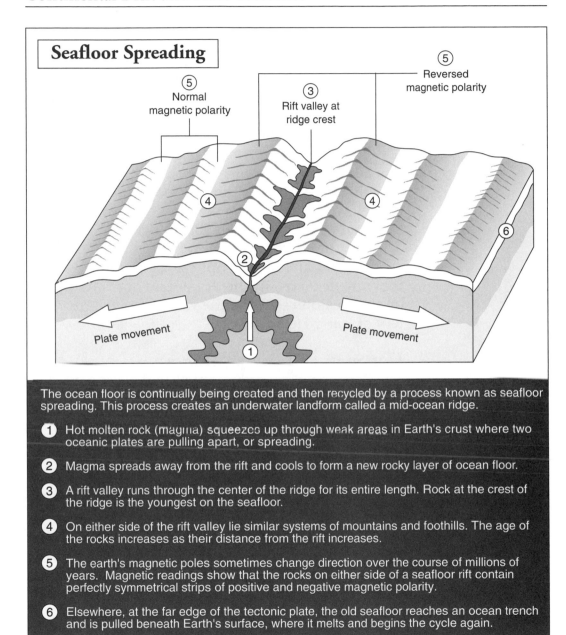

Seafloor Spreading

⑤ Normal magnetic polarity

③ Rift valley at ridge crest

⑤ Reversed magnetic polarity

④

④

⑥

② ①

Plate movement   Plate movement

The ocean floor is continually being created and then recycled by a process known as seafloor spreading. This process creates an underwater landform called a mid-ocean ridge.

① Hot molten rock (magma) squeezes up through weak areas in Earth's crust where two oceanic plates are pulling apart, or spreading.

② Magma spreads away from the rift and cools to form a new rocky layer of ocean floor.

③ A rift valley runs through the center of the ridge for its entire length. Rock at the crest of the ridge is the youngest on the seafloor.

④ On either side of the rift valley lie similar systems of mountains and foothills. The age of the rocks increases as their distance from the rift increases.

⑤ The earth's magnetic poles sometimes change direction over the course of millions of years. Magnetic readings show that the rocks on either side of a seafloor rift contain perfectly symmetrical strips of positive and negative magnetic polarity.

⑥ Elsewhere, at the far edge of the tectonic plate, the old seafloor reaches an ocean trench and is pulled beneath Earth's surface, where it melts and begins the cycle again.

reversals. Pipes thousands of feet long drilled deep into the sea bottom and brought up samples of rock and sediment referred to as plugs or cores. Specially outfitted ships conducted the difficult work that Ellen Prager, a marine scientist, compares to "drilling a hole in a New York sidewalk with a strand of spaghetti dangled at night, in swirling winds, from atop the Empire State Building."[16]

On board the ships or back in laboratories, the cores were analyzed and dated. Scientists found that rocks removed from the crest of a ridge were younger than those taken farther away. Rocks near the deep ocean trenches were the oldest on the ocean floor. The age of the rocks increased as their distance from the ridge increased. Hess, Vine, Matthews, and Morley were correct; the creation of new seafloor is balanced with its return to Earth's mantle.

With the confirmation of seafloor spreading, renowned Canadian geophysicist J. Tuzo Wilson took the next step in the development of the theory of plate tectonics. Along the way, he solved the mystery of Harry Hess's guyots.

Hot Spots and Paving Stones

The flat-topped mountains that had fascinated Hess also intrigued Wilson. He theorized that the guyots were connected with seafloor spreading through weak areas in Earth's crust called "hot spots." Wilson linked hot spots with the formation of guyots through this image: "Wilson imagined himself lying on his back in a stream, looking up toward the surface and breathing through a straw bent downstream by the force of the current. This led to his proposal that, as the Pacific seafloor crust drifts across a hot spot . . . magma erupts through to the surface, creating a linear chain of oceanic activities."[17]

Hess's guyots were part of an archipelago that had formed as the spreading Pacific seafloor passed over a hot spot. This chain of underwater volcanoes, islands, and flat-topped mountains stretches sixteen hundred miles from Hawaii through Midway Island to the Emperor Seamounts off of the Alaskan coast.

Wilson's greatest contribution to plate tectonics, however, was his vision of crustal plates. He proposed that the stiff crust of Earth is broken into moving blocks or plates that cover its surface like paving stones. The enormous plates are edged with either deep ocean trenches where two plates meet or with mid-ocean ridges, the regions where plates are moving apart. Wilson also solved the problem of flat plates moving over a round Earth by suggesting that fractures in the crust, called faults, cross the ocean ridges and allow the rigid plates to bend over a spherical surface. Wilson's model of interrelated blocks, faults, and trenches proposed a global system of plates that grind, collide, or tear apart in their journey across the globe.

A Global View

Plate tectonics offers a context for understanding and interpreting the planet's dynamic systems. From petroleum geologists who use the principles of plate tectonics to find oil reserves to seismology teams exploring a volcano's caldera, those who study Earth rely on plate tectonics to help them comprehend their world. Prager writes in her book *Furious Earth*, "We no longer call it continental drift or the plate tectonics theory, but simply plate tectonics. Few other scientific revelations can surpass plate tectonics in helping us understand the present, past and future shape of the Earth's surface, and its earth-shaking, fiery, and watery displays of power."[18] As scientists continue to discover more about Earth, the theory of plate tectonics changes and evolves. Even so, this theory based on Wegener's idea of moving continents, once dismissed as complete nonsense, has given scientists a lens for viewing the physical world.

Chapter 2

The Unified Theory of Plate Tectonics

The theory of plate tectonics describes how the movement of crustal plates has shaped the surface of Earth. As the plates collide, separate, and slide past each other, tectonic forces change the physical landscape: Over millions of years some regions uplift into mountains while others sink below sea level and fill with water. Tectonic forces also cause earthquakes and the formation and eruption of volcanoes. These events take place largely along the edge of Earth's crustal plates, the continent-size slabs of rock that astronauts have likened to pieces of an enormous jigsaw puzzle.

The Layers of Earth

Earth is a giant sphere made of five layers. At the center of Earth is a two-layered core: a hot, solid inner core surrounded by a liquid outer core, which is even hotter. Above the core lies a region called the mantle, composed of a lower layer and an upper layer. On the top of the mantle rides Earth's thin outer crust, broken into the tectonic plates. Since the crust is easily studied, more is known about it than the other layers.

Earth's outer crust is a solid, rigid layer of rock, but the crust of the continents is different from the crust of the ocean basins. Continental crust is made mostly of relatively light granite with an average thickness of twenty-five to thirty miles and includes some of the oldest rocks on Earth's surface. In comparison, ocean crust is made of basalt, a dense volcanic rock. The ocean basins have an average thickness of three miles and are less than 200 million years old. Granite is lighter than basalt, and these differences in the properties of crustal rocks help drive the tectonic forces of Earth.

Beneath the crust lies the mantle. It is about eighteen hundred miles thick and is made of red hot silicate rocks as well as iron and magnesium. While

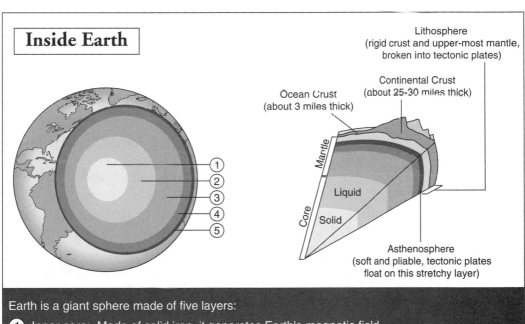

Inside Earth

Lithosphere
(rigid crust and upper-most mantle, broken into tectonic plates)

Continental Crust
(about 25-30 miles thick)

Ocean Crust
(about 3 miles thick)

Mantle

Liquid

Core

Solid

Asthenosphere
(soft and pliable, tectonic plates float on this stretchy layer)

Earth is a giant sphere made of five layers:

1 Inner core: Made of solid iron, it generates Earth's magnetic field.

2 Outer core: Very hot molten (liquid) iron. The combined inner and outer core are about forty-three hundred miles in diameter.

3 Mantle: This thickest layer is very dense and solid. Along with the upper mantle, this layer is approximately eighteen hundred miles thick. The entire mantle is composed of red hot silicate rocks as well as iron and magnesium.

4 Upper mantle: The thinner upper mantle is rigid and brittle.

5 Crust: A solid, rigid layer of rock, this is the only layer scientists are physically able to study.

scientists cannot yet drill to the mantle to retrieve core samples, they are able to study rock fragments from the mantle that were brought to the surface by volcanic eruptions.

The mantle has two layers, upper and lower. The thin upper mantle is rigid and brittle, while the lower mantle is very hot, dense, and solid. The two layers make up 82 percent of Earth's volume and 63 percent of its mass. The upper mantle and the crust of the planet combine to make up a region called the lithosphere, a strong, solid, rigid layer approximately sixty-two miles deep. Its cold, brittle surface is broken into moving pieces of crust called lithospheric or tectonic plates. These plates ride on the next layer, the asthenosphere, a transition zone thirty to eighty miles thick between the upper mantle and crust and the hot, solid lower mantle. The asthenosphere is soft and pliable like plastic and flows easily. It can be pulled and stretched and will flatten under its own weight. The hard tectonic plates float on the warm asthenosphere like bits of chocolate on warm honey.

Below the mantle, at Earth's center, lies the core, a ball of extremely dense iron about forty-three hundred miles in diameter. Its temperatures reach 10,800 degrees Fahrenheit, hotter than the surface of the sun. The inner core is a spinning mass of solid iron but the outer core is molten. The inner core is responsible for generating Earth's magnetic field: The kinetic energy from the spinning motion is converted into electrical and magnetic energy. The two-part structure of the core is like a natural electrical generator.

Looking Inside Earth

To learn more about the deep structures of Earth, scientists use a technique called seismic tomography to make three-dimensional computerized images of Earth's mantle. During the last fifty years scientists

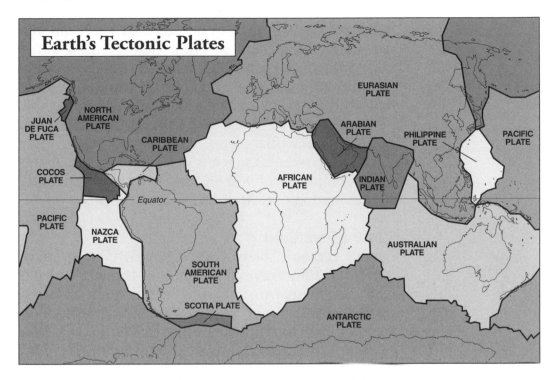

Earth's Tectonic Plates

EURASIAN PLATE

JUAN DE FUCA PLATE

NORTH AMERICAN PLATE

CARIBBEAN PLATE

ARABIAN PLATE

PHILIPPINE PLATE

PACIFIC PLATE

COCOS PLATE

AFRICAN PLATE

INDIAN PLATE

Equator

PACIFIC PLATE

NAZCA PLATE

AUSTRALIAN PLATE

SOUTH AMERICAN PLATE

SCOTIA PLATE

ANTARCTIC PLATE

have been able to produce a global view of Earth's internal structure by listening to its seismic events such as earthquakes beneath its crust. As rocks break and slip during earthquakes, the resulting seismic waves travel deep inside the planet and are recorded by sensitive equipment at monitoring stations all over the world. Seismic waves travel more slowly in hot, low-density rock and speed up in colder, denser regions.

Seismologists use computers to analyze the information from thousands of earthquakes. By mapping the changes in speed captured by many billions of bits of data, scientists are able to make pictures of Earth's interior. Andrew Alden, an author who writes about Earth, compares seismic tomography to "the doctor tapping on your chest and listening to the sounds to assess your internal state, only the Earth makes its own taps. After a century of collecting earthquakes, we're able to make some impressive maps of the mantle."[19]

It is now known that the lithosphere is cracked into seven major plates and several smaller ones.

The plates cover the surface of Earth like broken pavement. Each plate touches three or more other plates. The main plates are the Pacific, Eurasian, North American, South American, African, Australian, and Antarctic plates. The smaller plates are the Philippine, Arabian, Juan de Fuca, Cocos, Nazca, Caribbean, Indian, and Scotia plates. The Pacific plate is the largest. It covers 20 percent of Earth's surface and is made completely of oceanic crust. The other plates are a combination of continental and ocean crusts.

Plate Movement

Earth's tectonic plates are constantly in motion. Although they travel more slowly than fingernails grow, scientists can track plate movement by using satellites and lasers. Instruments aboard satellites measure how long it takes a beam of light to bounce from one continent on one plate to another continent on a different plate. By taking numerous measurements scientists are able to calculate the rate at which the plates are moving and the direction.

Plate movement continuously transforms the surface of Earth. The slow, steady travels of tectonic plates cause mountains to rise, basins to sink, and ocean boundaries to change. Geophysicists Allesandro Forte of the University of Western Ontario and Jerry Mitrovica of the University of Toronto are amazed that forces deep within Earth can shape its surface: "It really ignites the imagination to realize how things are changing hundreds of [miles] kilometers beneath your feet and how this change connects to majestic features on the Earth's surface."[20] Without tectonic forces, erosion would wear down the tallest peaks and fill in the deepest lakes. Plate movement prevents the surface of Earth from becoming flat like the moon's.

Plate Boundaries

The world's tectonic plates move by pulling apart, by colliding, and by sliding past each other. Divergent boundaries are zones where plates pull apart, while convergent plate boundaries are regions at which plates have collided, forcing one slab under another. Areas where the plates move, or attempt to move, horizontally past each other are called transform boundaries.

Divergent boundaries are spreading centers. As Earth's crust tears and causes two neighboring plates to pull apart, underwater ridges form and deep rifts open. Most divergent regions lie deep under the ocean but some are on land.

Africa's Eastern Rift Valley is an example of a terrestrial divergent zone. As the Somalian plate pulls away from the African plate, the continental crust stretches and cracks at Earth's surface. Magma squeezes through the rifts and sometimes erupts into volcanoes like Mount Kilimanjaro, Africa's largest volcano. The pressure from the rising magma also creates more fractures or faults. If the plates continue to move apart, the Indian Ocean may flood the area, creating a new ocean and turning the Horn of Africa into a large island.

Another example of a land rift is the island of Iceland located between Greenland and the United Kingdom. Iceland, slightly smaller than Kentucky, straddles the mid-Atlantic ridge where the North American plate is pulling away from the western part of the Eurasian plate. Two hundred volcanoes and 250 hot springs cover the little island. Tectonic forces are splitting Iceland as the North American plate continues to move westward. Tectonic forces also shape the ocean floors.

A Growing Ocean

The Atlantic Ocean formed when the North American plate split from the Eurasian plate. Today the North American plate continues to move away from Eurasia in a southwesterly direction at a rate of one half of an inch in a year. The Eurasian plate travels

A team of hikers is dwarfed by the deep rift that cuts across Iceland. Tectonic forces are splitting Iceland in two.

one half of an inch a year in an east-southeasterly direction. Because of the plate's movements, colonists who sailed from Europe in the 1600s to the New World crossed a smaller Atlantic Ocean. The North American and European continents are now thirty feet farther apart. As the Atlantic Ocean is getting bigger, however, the Pacific Ocean is shrinking: Its plate is relentlessly colliding with the Eurasian plate along a convergent boundary.

Convergent zones are regions where tectonic plates ram into each other. More tectonic activity takes place at convergent boundaries than at any other boundary: Eighty percent of the world's volcanoes, 90 percent of all earthquakes, and most of the world's mountain building occur along convergent boundaries. Tectonic plates converge in three ways: An oceanic plate may dive beneath a continental plate, two continental plates may smash together, or two oceanic plates may meet head-on with the older, colder plate sinking below the younger plate. Old crust is destroyed as two plates meet, but each year tectonic activity along the ridges creates about three to five cubic miles of new oceanic crust. Since this amount is nearly equal to the volume of old crust recycled annually into the mantle along convergent plate boundaries, the overall size of Earth stays the same.

Deep Sea Trenches

The collision of oceanic plates often forms deep-sea trenches and island arcs parallel to the trenches. When two oceanic plates meet, the heavier, older plate plunges beneath the less dense one. As the denser plate sinks into the soft asthenosphere, a deep trench appears along the margins.

The Mariana Trench in the South Pacific was created by the subduction of the faster-moving Pacific plate beneath the smaller Philippine plate. The Mariana Trench stretches for 1,580 miles in the South Pacific and contains the deepest point in the world's oceans, the Challenger Deep. More than six times deeper than the Grand Canyon, the Challenger Deep plummets to a depth of 35,838 feet. In fact, if a mountain as tall as Mount Everest rose up from the floor of the Challenger Deep, the mountain's top would still be one mile below the surface of the ocean. At a depth of nearly seven miles, the water pressure is about eight tons per square inch, or the equivalent of balancing four cars on an adult-size thumb.

Island arcs often run parallel to the deep ocean trenches. The Aleutian Islands of Alaska as well as the islands of Japan and the Philippines were formed by the subduction of the Pacific plate. Rising magma erupted into underwater volcanoes that eventually became island chains.

When Oceans and Continents Collide

Subduction is one result of the difference between oceanic crust and continental crust. Basaltic ocean crust is colder, denser, and heavier than the granitic crust of continents. Along their convergent boundary, the heavier oceanic plate plunges beneath the lighter continental plate. Eventually the subducted plate sinks into the asthenosphere and gradually disappears into the mantle. While deep trenches form in the ocean basins, volcanoes and mountains are thrust up on land as the lower plate bends downward.

Volcanoes form in Earth's lower mantle where temperatures are hot enough to melt rock into the flowing form called magma. Since it is less dense than the solid rock of the upper mantle, magma starts to travel up toward the crust, pushing its way through cracks in the mantle. As the magma gets closer to the surface, gases begin to form, rising faster than the magma. If a vent or fissure is present at the surface, the gases and magma are forced out explosively through it in a volcanic eruption. Once above Earth's surface, magma is called lava.

Shield volcanoes erupt slowly and quietly and are formed as lava oozes out of cracks and fractures in Earth's crust. Shield volcanoes have gently sloping sides made by layers of viscous, or runny, lava, and are wider than they are tall. For millions of years, eruptions added layer upon layer of lava to build up Mauna Loa, a shield volcano in Hawaii. This enormous volcano is the world's largest, and if measured

Plate Boundary Types

Divergent Boundaries (Spreading Plates)

Deep rifts open and molten rock squeezes up and out. Most divergent regions lie under the ocean.

Convergent Boundaries (Colliding Plates)

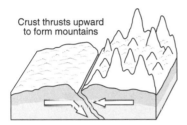

Crust thrusts upward
to form mountains

Continental–Continental Boundary

The leading edges of two continental plates
buckle and fold as they crash into one another.

Volcanoes often form on
continental plate near boundary

Oceanic–Continental Boundary

The colder, denser oceanic crust sinks beneath the lighter
continental plate and gradually disappears into the mantle.

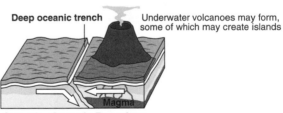

Deep oceanic trench

Underwater volcanoes may form,
some of which may create islands

Oceanic–Oceanic Boundary

The heavier, older plate plunges beneath the younger one, creating a deep trench along the margins.

Transform Boundaries (Sliding Plates)

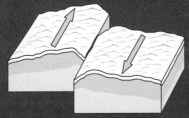

Crustal plates grind past one another, sometimes in sharp jolts felt as earthquakes.

On May 18, 1980, Mount Saint Helens explodes in a violent eruption, shooting ash, steam, and huge chunks of rock thousands of feet into the sky.

from its base deep below the Pacific Ocean, it is also the world's tallest mountain, though most of it lies below the waves.

Stratovolcanoes, or composite volcanoes, explode violently and can cause enormous destruction to the surrounding landscape. Oceanographer Ellen Prager compares stratovolcanoes to "huge bottles of champagne. . . . Just as the champagne cork can fly violently through the air, so can volcanic rocks, hunks of lava, ash and gas."[21]

These large symmetrical volcanoes are made from layers of lava and avalanches of hot ash, bits of rock, and gases called pyroclastic flows that can race down the slopes of a volcano at speeds too high to permit escape by car or train. They bury trees, buildings, and even towns in their path. In the late twentieth century Mount Saint Helens, a large stratovolcano in the western United States, erupted with a violence that sent a deadly mix of rocks, ash, and gas surging down its slopes at an incredible rate of three hundred miles per hour.

Mount Saint Helens

To the west of the Oregon and Washington coasts, a small oceanic plate called Juan de Fuca is sinking below the westerly moving North American plate. As the Juan de Fuca plate continues to move beneath the continental plate, its crustal material begins to melt. Magma from the melting plate makes its way to the surface approximately three hundred miles inland, forming the Cascade mountain range and its volcanic peaks.

Mount Saint Helens, Mount Rainier, and Mount Hood are active volcanoes in the Cascade Range. On May 18, 1980, Mount Saint Helens erupted with a ferocity that shot a cloud of superheated ash and gas fifteen miles high into the atmosphere. The explosion killed sixty people, destroyed 150 square miles of forest, and buried river valleys and towns beneath layers of ash, mud, and debris.

Randall Lovett lived in Hazel Dell, Oregon, about thirty miles from Mount Saint Helens. He describes the eruption as "a magnificent sight. It was like an atom bomb going off on the top of St. Helens. I also noticed what I thought were rocks coming out of the top of the plume. Later we read that these rocks were actually chunks of earth the size of a house, being thrown 60,000 feet in the air."[22] Mount Saint Helens and the Cascades are a part of the world's most active tectonic region, the Pacific plate's Ring of Fire.

The Ring of Fire and Its Geological Offspring

The Ring of Fire is a belt of volcanically active zones that surrounds the enormous Pacific plate and includes

75 percent of the world's active volcanoes, or nearly six hundred peaks. As the Pacific plate moves counter-clockwise at 2.5 inches per year, it slides beneath its neighboring plates and begins to melt. The resulting magma escapes through gaps and fractures in the surface to form a ring of volcanoes around the plate's edge.

These volcanic zones are home to some of the world's most violent eruptions. Krakatoa, an island between Sumatra and Java in the western South Pacific Ocean, exploded in 1883. The eruption generated giant waves, called tsunamis, that killed over thirty-six thousand people. The eastern edges of the Pacific plate are also home to destructive volcanoes. In 1985 Nevado del Ruiz, a volcano in the Andes Mountains of Colombia, erupted and set off massive mudflows that killed twenty-five thousand people. The same tectonic forces that fed the blast of Nevado del Ruiz also had created the mountain chain of which that volcano is a part.

The Andes Mountains of the western coast of South America run along the convergent boundary of the Nazca and South American plates. As the heavier Nazca plate dives beneath the continental crust of the South American plate, the lighter plate crumples and uplifts into jagged mountain ranges. Magma from the melting Nazca plate also rises through the lithosphere to form volcanoes along the backbone of South America. While the Andes Mountains were and continue to be created by oceanic-continental collisions, some of the world's tallest mountain ranges are formed as the edges of colliding continental plates are thrust miles into the atmosphere.

Continental Collisions

The clash of continents results in spectacular landforms. Continental crust is light and buoyant and does not subduct into the mantle. Therefore, the convergence of continents is like the collision of icebergs—neither sinks. Instead, the leading edges of the massive slabs of granite buckle and fold into steep mountains in long parallel ranges. For example, the Himalayan Mountains, the highest in the world, were shoved up when the swift-moving Indian plate smashed into the Eurasian plate.

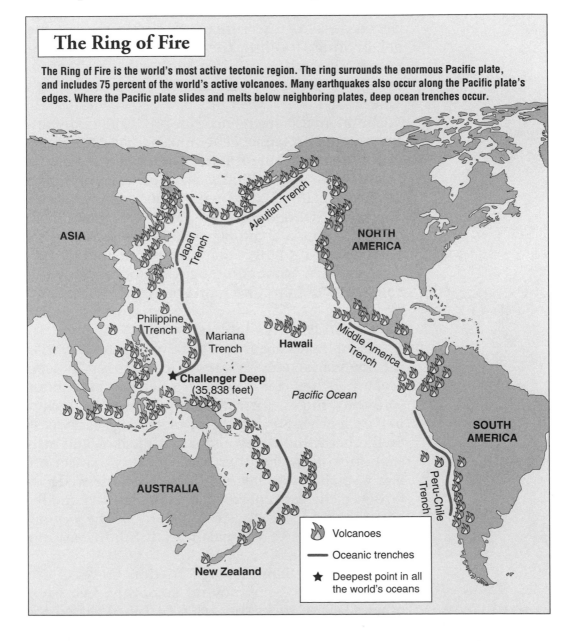

The Ring of Fire

The Ring of Fire is the world's most active tectonic region. The ring surrounds the enormous Pacific plate, and includes 75 percent of the world's active volcanoes. Many earthquakes also occur along the Pacific plate's edges. Where the Pacific plate slides and melts below neighboring plates, deep ocean trenches occur.

ASIA

NORTH AMERICA

Aleutian Trench

Japan Trench

Philippine Trench

Mariana Trench

Hawaii

Middle America Trench

★ Challenger Deep (35,838 feet)

Pacific Ocean

SOUTH AMERICA

AUSTRALIA

Peru-Chile Trench

New Zealand

🔥 Volcanoes

— Oceanic trenches

★ Deepest point in all the world's oceans

About 60 million years ago the Himalayan Mountains along the Tibet and Nepal borders rose up when the Indian plate charged across the equator and rammed into the Eurasian plate. Since both plates have about the same rock density, neither was forced beneath the other plate. The force of the collision crushed the plates' borders into the steep, jagged peaks of the Himalayas and thrust up the neighboring land into the Tibetan Plateau. Mount Everest, the highest point on Earth with an elevation of 29,035 feet above sea level, shot up during this time.

Continental convergence has produced many ancient mountain ranges. The Appalachian Mountains in the eastern United States rose to great heights about 440 to 480 million years ago when the North American plate collided with other ancient plates. The Alps of Europe were born as Africa crashed into southern Eurasia about 80 to 100 million years ago. These mountain ranges mark the boundaries of convergent zones. Mountain ranges also occur along the boundaries of transform fault zones.

Earthquakes and Plate Tectonics

Transform faults are areas where two plates meet. At the boundaries, crust is not destroyed as it is in deep ocean trenches, nor is it created like it is at seafloor spreading ridges. Instead, the crustal plates grind past each other. Some tectonic plates edge easily past each other without any major earthquakes, and others hop past one another in small jolts that set off many small earthquakes. If plates become wedged, however, the earthquakes caused when they finally break loose can be disastrous. Many of the world's most devastating earthquakes result from transform fault movement.

Earthquakes happen when friction prevents the tectonic plates from moving smoothly past each other. Slabs of rocks that have become wedged together now may begin to bend and twist. Even though the rock is solid and rigid, under certain conditions it can stretch like a rubber band and store up energy: "Rock is squishy on very long time scales,"[23] points out Andrea Donnellen, a geophysicist at NASA's Jet Propulsion Laboratory. Eventually the rocks along both sides of the fault shatter, and like an overstretched rubber band, their built-up energy is released in the form of an earthquake.

As seismic waves spread out from an earthquake center, Earth's crust trembles. Geophysicist Lucy Jones of the U.S. Geological Survey explains earthquakes: "It's just like when I snap my fingers. I release energy that's making the air vibrate. When a fault slips, it releases energy that makes the ground vibrate. It's that secondary effect, that shaking, that we perceive as an earthquake."[24]

Thousands of earthquakes occur each year. Most are minor earthquakes that are too small to be felt. Some, however, are intense. Transform faults, in particular, are capable of producing sudden and violent earthquakes.

Faults, Tsunamis, and Hot Spots
The San Andreas Fault is the world's most visible transform fault. It lies between the North American and Pacific plates and stretches over a thousand miles from northern California through the western coast to the Gulf of California in northwestern Mexico. As the two plates grind past each other, the sections along the fault jam and pressure builds up. The strain is relieved by minor tremors; in a quiet year in California, ten thousand small earthquakes are recorded. Sometimes, however, the plates jolt abruptly, producing violent seismic events; several catastrophic earthquakes have hit the state in the past two hundred years.

Sudden fault movement may also produce tsunamis, the Japanese term for "great harbor waves." A

The world's most visible transform fault, California's San Andreas Fault, marks the boundary where two plates grind and slip past one another.

tsunami can be created by a sudden movement on the seafloor, an underwater explosion, or the impact of a large object falling into the sea such as a land-slide. Each of these events displaces large amounts of water. These fast-moving, low, long waves spread out in all directions like ripples spread out when a pebble is dropped in a pond. While some tsunamis travel short distances, others race thousands of miles across the open ocean at rates of five hundred miles per hour, or nearly the cruising speed of a Boeing 737 jet. Eventually the waves reach land. As they approach the coasts, the waves bunch up and increase in height.

Deep in the center of the Pacific plate, nearly six thousand miles from the nearest plate border, Hawaii has been swept by numerous tsunamis. An anomaly on the Pacific plate, Hawaii was not formed by either colliding or spreading tectonic plates. Instead, the island chain grew over a hot spot.

Hot spots are regions where rising columns of magma push through weak spots in the crust of tectonic plates. These columns, or plumes, originate in the mantle deep below the plates. Millions of years ago, magma escaped from the mantle and rose through the asthenosphere. As the magma plume pushed against the bottom of a lithospheric plate, it stretched and cracked the surface crust. Eventually, magma was able to force its way through the cracks to the surface and form volcanoes or areas of volcanic activity.

Hot spots can be found beneath both continents and oceans. The islands of Hawaii are the end of a long chain of islands and submerged sea mounts that were created as the enormous Pacific plate passed over an ancient hot spot. In the western United States, the volcanoes and lava flows of Arizona's San Francisco Volcanic Field were created by a hot spot that lies under the North American plate.

Chapter 3

Plate Tectonics Today

Technology is transforming the picture of Earth. Deep-sea submersibles, or small submarines capable of withstanding immense pressures, carry scientists miles below the waves to the bottom of deep-sea trenches. For months of uninterrupted time, small autonomous robots explore canyons and cracks in the ocean floor. Satellites provide three-dimensional views of Earth. Complex computerized models are capable of compressing tectonic processes that take millions of years into a matter of minutes. From simple radio receivers to satellite-mounted lasers, the advances of technology are also advancing the theory of plate tectonics.

GPS, the Global Positioning System

One way to measure the movement of the tectonic plates is the use of the Global Positioning System, or GPS. GPS allows earth scientists to monitor changes in direction, speed, and altitude of the plates. With this information they are better able to forecast natural disasters such as earthquakes and volcanic eruptions. Observations of crustal changes also give geologists and geophysicists a better understanding of plate tectonics by allowing them to survey and map the planet's crust with great accuracy since GPS units can track plate movements of less than a quarter of an inch. By combining earthbound receivers and orbiting satellites, the Global Positioning System is revolutionizing how scientists study Earth.

The Global Positioning System relies on a network of twenty-seven satellites in orbit around Earth. Each satellite completes two orbits around Earth every day. Continuous information about the latitude, longitude, and elevation of the GPS satellites is broadcast on radio waves to receivers on the ground. With signals from at least four satellites, the GPS receivers can calculate the location on the planet's surface of the phenomenon the signals are pinpointing. Common GPS receivers used in boats and cars are accurate up to twenty or thirty feet but receivers used by earth scientists can give locations to within an eighth of an inch. These advanced units let scientists track the fine plate movements found in earthquake-prone areas or on volcanoes.

A geologist sets up a GPS station near the top of Mount Saint Helens. GPS devices help geologists track plate movements on volcanoes and in earthquake-prone areas.

Vulcanologists, scientists who study volcanoes, and geodesists, scientists who measure Earth, utilize GPS stations to observe the changes in the shape of a volcano. Even slight changes can herald a volcanic eruption. Small bulges or swellings appear on the surface of the volcano as magma approaches the surface. GPS stations are placed on active volcanoes to determine the direction and rate of these tiny surface movements. The measurements help scientists to estimate the location, depth, and amount of the rising magma. With continuous GPS monitoring, earth scientists have improved the capability to warn of possible volcanic eruptions. GPS is also used to predict earthquakes.

After the devastating 1994 earthquake near Northridge, California, a sophisticated network of 250 GPS receiving stations was set up throughout

A geologist uses a system of laser beams to detect small movements of the ground near Parkfield, California.

Southern California, a complex region of shattered plates and faults. The process behind GPS monitoring, however, is relatively simple. NASA's Andrea Donnellen explains, "With a GPS network, we are essentially just measuring the location of points. So we can take two points on the Earth, and we see how these points get closer together or farther apart."[25] With information from the network, scientists evaluate where strain is building up along the fault zones. By being able to locate strain in faults, scientists hope to create timely and accurate earthquake forecasts. The U.S. Geological Survey (USGS) has placed earthquake monitoring stations in the hopes of catching the next "Big One" before it strikes California.

The Parkfield Earthquake Experiment

The U.S. Geological Survey measures crustal change across the United States, especially in the western regions of the country, where most earthquakes occur. During the 1980s the USGS and the California Geological Survey began the Parkfield Earthquake Experiment to monitor the movement of the San Andreas Fault. Parkfield, California, is located along the section of fault where small earthquakes shake the ground regularly. Seismologists use many instruments, some very sophisticated, some quite simple, to check for changes in stress and motion. Then they analyze the data for danger signs to warn residents of a possible earthquake.

While no major earthquake has occurred in the Parkfield area, scientists have gathered significant information about plate movement along the fault zone. Scientists like Nancy King of the USGS believe the value in earthquake research lies in its potential to help people: "We're trying to figure out why earthquakes occur and how they occur. We're doing this partly because it's an interesting scientific problem, but the main reason we're doing it is because it has societal value. If we can understand earthquakes and figure out which areas are subject to seismic hazard, we hope that we can save lives and property."[26]

By analyzing the information obtained from the Parkfield Earthquake Experiment scientists have gained a detailed view of local geological events. Other programs, however, have provided a large-scale picture of the tectonics forces. One such project, Canada's *LITHOPROBE*, highlights technologies that have created a comprehensive view of continental tectonics.

Continental Tectonics and the *LITHOPROBE* Project

LITHOPROBE is a multidisciplinary earth science study of the North American continent and its oceanic margins. Research teams of academic, government, and resource industry earth scientists explore Canada's rich geological history of the last four billion years. By studying Canada's geological evolution, *LITHOPROBE* offers a window onto the tectonic forces that also shaped the entire world.

LITHOPROBE's geologists and geophysicists are piecing together a complete picture of the planet's history. Ray Price, professor of geological sciences at Queen's University in Canada, explains that the project does more than provide scientists with a three-dimensional view of Earth. It also adds a fourth dimension—time. Dr. Price explains:

On Mars, the moon, Venus, the evidence of what happened at their beginnings is still there: the accretion of extra-terrestrial scars, the craters. The Earth is different; it's an active planet, its history pretty much destroyed by continents drifting, mountain-building, erosion. Lithoprobe now bridges the gap between geology and planetology. We get the whole history of the Earth, back to when the planet was forming, back to the beginning of time.[27]

LITHOPROBE explores the depths of the planet's lithosphere through seismic surveys in which sound waves are generated in one location and picked up in a different one. The survey lines can be hundreds of miles long, so special trucks were designed for the project. These vibroseis trucks, as the unique vehicles are technically known, are capable of driving on highways yet they also carry the heavy equipment that produces the seismic waves for the surveys. As people watched the vibroseis trucks in action, they became popularly known as dancing elephants.

This giant vibroseis truck can be used to send sound waves deep into the earth in order to create models of rock structures up to sixty miles below the surface.

Dancing Elephants

The *LITHOPROBE* project has twelve of the huge vibroseis trucks. Each one has a hydraulic plunger that
lifts the forty-four-thousand-pound truck off the
ground. Then, the hydraulic system shakes the
truck. As the "elephant" seems to dance in the air,
sound waves are pumped into the rock below.
Echoes of the waves are picked up by sensors. Since
sound waves travel at different rates through different kinds of rock, geologists can create images of the
rock boundaries. They can also find geological structures such as ancient faults.

LITHOPROBE's dancing elephants have surveyed
ten horizontal transects, or lines, across Canada
and parts of the United States. Computers analyze
the survey information and generate pictures of
the lithosphere up to sixty miles deep. Gravi-
tational and magnetic measurements are added, and
four-dimensional models are generated that show
how Canada changed over billions of years. Richard
Wardle, a senior geologist with the Newfoundland
Geological Surveys Branch, believes the project has allowed scientists to peer deep below the earth and back
into time.

> What we've learned is what actually happens
> when one continent collides with another and
> all the bits and pieces get squeezed together.
> We've found rocks that originally formed at the
> surface and have been plunged down 30 to 35
> kilometers into the lithosphere, almost to the
> base of the crust. We've seen volcanic offshore
> islands that got squished between two colliding
> continents. We've seen how Labrador and
> Greenland split 80 million years ago. We've
> seen rocks here that are the same as rocks in
> Scandinavia—from the time the two continents
> were one. The history of the Earth is one of con-
> stant collision and separation.[28]

While earth scientists like those who run *LITHO-PROBE* examine the geology of continents, others take to the seas. They brave the crushing pressures, the impenetrable darkness, and the numbing cold of the deep ocean to witness firsthand the forces that shape the planet. With less than one percent of Earth's ocean mapped, much remains a mystery. The next generation of geologists, oceanographers, submersible pilots, robot designers, and marine engineers are increasing that percentage.

Trieste and *Alvin,* the Submersibles

In 1960 a small underwater vehicle, the submersible *Trieste*, took two men down to the deepest spot in the world, the bottom of the Mariana Trench's Challenger Deep in the Pacific Ocean. It was the first and only time humans reached a depth of 35,797 feet. Since then scientists have continued to explore the abysses of the ocean, often in somewhat larger submersibles crowded with men, equipment, and motors.

Submersibles are small, manned vehicles that are designed to explore the ocean floor. *Alvin* is one of the best known deep-sea submersibles. Owned by the U.S. Navy and operated by Woods Hole Oceanographic Institution in Massachusetts, the titanium-hulled *Alvin* can carry three passengers down to 14,764 feet, or more than 2.5 miles deep. Over forty years the little sub completed about four thousand dives, found a hydrogen bomb, explored the sunken Titanic, discovered amazing life forms that draw their energy from chemicals, not sunlight, and confirmed that the seafloor is spreading. "The contribution *Alvin* has made to science is unquestioned,"[29] says Robert Gagosian, president and director of Woods Hole, commenting on the submersible's part in unlocking the mysteries of the ocean.

For the scores of scientists who have traveled to the bottom of the sea in *Alvin,* the experience is

For forty years, the deep-sea submersible Alvin *has helped scientists learn more about ridges and rifts on the ocean floor.*

unforgettable. Brad Stevens, a biologist with National Marine Fisheries Service in Kodiak, Alaska, describes reaching the ocean floor as "landing on the moon but you're landing on Earth—a part of Earth you've never seen before, that nobody's ever seen."[30] Cindy Lee Van Dover, a deep sea biologist and submersible pilot, also vividly remembers her trips in *Alvin*.

Van Dover piloted *Alvin* for two and a half years. In the small sub, she has explored a tiny fraction of the mid-ocean ridges and crawled along their rift valleys. She paints a dramatic picture of the depths: "The tortured terrain in the valley is filled with pits, caverns, tall pillars, and rubble. Lavas frozen in motion form pillows, whirlpools and ripples, drapes and lobes. It is a submarine landscape of stark beauty."[31]

At thousands of feet deep, she has also come face-to-face with the enormous forces that shape the planet.

> I have sometimes found fissures in these valleys large enough to descend into with the submersible. The walls of a fissure can be so sheer that the seafloor looks as if it had been sliced and wedged apart by a giant knife. I have driven deep into a fissure where the walls narrowed to within a yard of *Alvin's* side windows. In front of me I could see the seam where the walls joined. I was at the very place where the planet's crustal plates were inexorably moving apart. I knew that hundreds of microearthquakes were being reported in the area—evidence that the plates were spreading—and I half-expected the fissure to open up and engulf me. I sought no excuse to loiter there.[32]

In August 2004 Woods Hole Oceanographic Institution announced that the sub would be retired soon and be replaced by a high-tech submersible that would be capable of diving to 21,320 feet. Still, *Alvin*, "the little sub that could," will be remembered for opening up the dark worlds at the bottom of the sea.

Jason, the ROV

Along with *Alvin*, Woods Hole Oceanographic Institution operates another famous underwater explorer, a little ROV called *Jason*. Underwater ROVs, or remotely operated vehicles, are small, tethered, unmanned submersibles. Capable of entering tight spaces and exploring dangerous areas, these little robotic vehicles have added an exciting element to deep-sea research. Cables or tethers tie the ROVs to a support ship on the surface. The boxy subs are "flown" by a human operator with a joystick in a control room aboard the ship. High-quality still and video cameras allow the onboard scientists and engineers to see remote areas of the ocean floor without any of the danger. ROVs are also equipped with sensors, mechanical arms, and sampling baskets. *Jason* was specifically designed to carry out scientific operations.

Jason, together with its partner ROV, *Medea*, has sampled hydrothermal vents, explored shipwrecks,

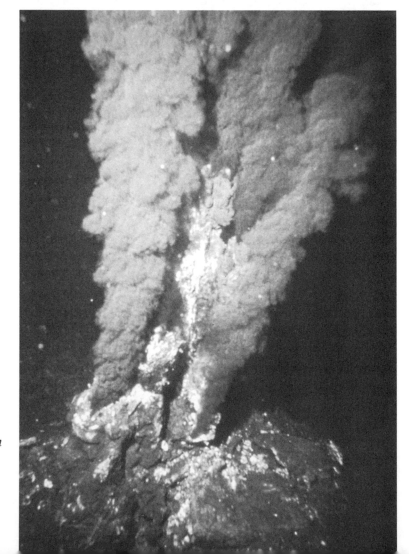

Hot water and mineral particles form a dark cloud as they spew from a hydro-thermal vent along the mid-ocean ridge.

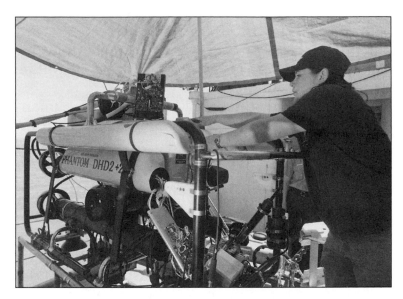

A scientist inspects an underwater remotely operated vehicle before deploying it into the water. ROVs can dive much deeper than manned submersibles.

and set up ocean floor observatories. Although only eleven feet long, eight feet high, and seven feet wide, *Jason* can dive to twenty-one thousand feet, much deeper than *Alvin*. Three miles of cable connect *Jason* and *Medea* to their support ship, the *Atlantis*, which is equipped with sophisticated research equipment and laboratories. Though *Jason* has the ability to obtain detailed images and specific data from the bottom of the ocean, it is limited by its tether and battery life. Eventually an ROV must return to its mother ship. Autonomous underwater vehicles, or AUVs, on the other hand, can stay down at depth for months, untethered and unsupervised.

ABE, the AUV

ABE is Woods Hole's AUV. *ABE* stands for Autonomous Benthic Explorer. *ABE* is not tethered to a ship; it is a true robot, flying free along the ocean floor unguided by human hands. An underwater robot is programmed to follow a predetermined path. Along the way, the little robot demonstrates its autonomy by mapping the seafloor with sonar, taking photographs, measuring the magnetism of rocks, and collecting samples all by itself. Once its mission is finished, the robot either "goes to sleep" to conserve its batteries and stays at the bottom for months, or it returns to the surface where it will be retrieved by its support ship.

AUVs have a distinctive advantage over human-operated vehicles. The underwater robots perform their assigned jobs precisely and tirelessly. "If we program a vehicle like ABE to fly in a grid pattern over the sea floor, it does exactly what we've asked and it doesn't get tired, it doesn't get bored, so it produces very high quality data,"[33] says Dana Yoerger, a Woods Hole scientist. That does not mean, though, that AUVs will replace people. They are a scientific tool that lets scientists further explore the depths of the ocean. AUVs will be a significant part of an ambitious project to study the ocean. The NEPTUNE Project will use AUVs to examine up close an entire tectonic plate.

NEPTUNE, an Underwater Observatory

The NEPTUNE Project, or the North-East Pacific Time-Series Undersea Networked Experiments, brings together a diverse group of nations, agencies, universities, and scientists to study the Juan de Fuca plate. The Juan de Fuca plate lies off the coast of Canada, the Pacific Northwest, and California. The project will turn the ocean floor into an underwater observatory where scientists can observe biological, chemical, and tectonic processes. John R. Delaney, NEPTUNE's program director, believes it will change how scientists work. "In 20 years, we'll turn the seafloor into a 'laboratory' where we'll routinely do work that isn't imaginable today even in labs on land."[34]

To create its underwater laboratory, NEPTUNE will "wire" the entire Juan de Fuca plate. Though tiny in comparison to the other tectonic plates, Juan de Fuca presents an enormous research area—about 150,000 square miles, or the combined size of

Indiana, Illinois, and Iowa. An eighteen-hundred-mile submarine network of fiber optics and cables will crisscross the plate. About a mile below the ocean's surface, thousands of instruments, sensors, cameras, and robotic vehicles such as ROVs and AUVs will plug into nodes, or experimental sites along the network. The nodes will distribute power to the instruments and robots. They will also provide real-time transmission of data and two-way communications. Scientists will be able to control underwater ROVs from labs hundreds of miles away on dry land.

The public will also be able to access NEPTUNE's wealth of images, live videos, and information via the Internet. Delaney believes this level of accessibility is an important goal for NEPTUNE. "Oceans are the most fascinating feature of our solar system," explains Delaney. "We want everyone with a home computer to be able to access what we're studying, and eventually involve school children in our robotic operations, allowing them to experience firsthand the mysteries of the deep."[35] The principal goal of the NEPTUNE Project, however, is to gain a clearer view of the Juan de Fuca plate, which provides a glimpse of geological forces at work. Earthquakes and volcanic eruptions occur along these zones as the smaller oceanic plate subducts under the larger continental North American plate. Through NEPTUNE's network of sensors and cameras, earth scientists will have instant and direct access to images and information of tectonic events.

The study of the ocean floor leads to an understanding of the forces that sculpt the planet's surface both under the water and on land. But to understand more about tectonic processes, scientists must go deep underground. One way to do this is to drill directly into Earth's lithosphere and retrieve core samples.

The Resolution *is a specialized vessel capable of drilling holes deep into the seafloor in order to retrieve core samples for study.*

Oceanic Drilling

Ocean drilling provided the first evidence of plate tectonics during the 1960s. Core samples taken by a research enterprise known as the Deep Sea Drilling Project, funded by the U.S. government, showed that ocean basins are relatively young. Analysis of the rock samples also confirmed the theory of seafloor spreading. Drilling is still an important tool today. It allows scientists to reach parts of Earth never seen before.

Starting in 1968 with the *Glomar Challenger* and the *Resolution* in the 1970s, specialized vessels called drilling ships have conducted investigations of the ocean basins by drilling deep into the ocean floor. Over twenty years, the *Resolution* has drilled seventeen hundred holes throughout the world's oceans, including continental margins and ocean trenches, and has added to the understanding of earthquakes, climate change, and natural resources.

The *Resolution* is a ship with a derrick that stands 202 feet above the waterline. Afloat in nearly twenty thousand feet of water, the ship can drill a hole over six thousand feet deep into the seafloor. Long sections of hollow drill pipe are threaded together and fitted with a drill bit at the end. This assembly is lowered to the seafloor, and the sharp teeth of the bit chew into the ocean crust. As the pipe turns it pushes down into the floor, forcing the rocks and sediments into the pipe where they form a long cylindrical sample called a core. The core is pulled out of the pipe on the ship to be studied. At times seismic instruments are lowered into the hole as part of a global seismic network project. In 2006 the *Resolution* will be joined by a new drilling ship sponsored by the Japanese government and called the *Chikyu*, Japanese for "Earth." The *Chikyu* is a massive ship and will be able to drill twenty-three thousand feet below the seafloor in water depths up to thirteen thousand feet.

Deep-sea drilling is one of the older methods of studying tectonic forces. Now earth scientists are able to reach out past the planet and search for these forces on other bodies in the solar system, using technologies from GPS to advanced radar systems to extravehicular devices borne into space by NASA orbiters.

Extraterrestrial Tectonics

Planetary geologists study geological processes on other planets and bodies in the solar system. Telescopes, both on the ground and in orbit around Earth, and spacecraft equipped with powerful cameras have given glimpses of the geological forces at work on the other planets. From these sources it appears that

Earth may be the only body in the solar system that is still tectonically active. On this planet volcanoes erupt, huge seismic tremors shake the earth, and immense slabs of crust slide past and under each other. But on other planetary bodies evidence of plate tectonics is either unclear or missing.

Tectonic activity needs an internal heat source. As heat escapes a planet's or moon's interior, it fuels plate tectonics. Smaller bodies lose heat more quickly. Mars, the moon, and probably Mercury, though once volcanically active, are now cold and lifeless. Planetary geologists have looked often at Earth's two neighbors, Venus and Mars, for signs of tectonic forces.

In 1990 the NASA *Magellan* spacecraft orbited Venus and surveyed the planet by means of an advanced imaging technique known as SAR for synthetic aperture radar. SAR emits radar waves able to penetrate Venus's thick atmosphere and bounce off land features on the planet's surface. From the *Magellan* mission and many others, planetary geologists determined that while Venus has more volcanoes than any other planet in the solar system, it did not develop plate tectonics. Mars, however, seems to have started forming plates.

Tectonics and Mars

From small probes to surface rovers, a wide variety of spacecraft has explored the red planet. The surface of Mars has been studied with such intensity that some scientists feel the home planet, Earth, is often overlooked. "We know more about Mars than we know about the oceans,"[36] objects Sylvia Earl, a former chief scientist with the U.S. National Oceanographic and Atmospheric Administration. These space missions, though, have yielded a wealth of information about Martian geology. Because there are giant rift valleys on Mars, planetary geologists believe plate tectonics started on the planet but that Mars cooled

too quickly, and the process halted. There are, however, other tectonically active bodies in the solar system.

Images taken by NASA's *Voyager* and *Galileo* spacecrafts show that Io and Ganymede, two of Jupiter's moons, are geologically alive. The 1979 *Voyager* mission to Io showed a moon devastated by volcanism; at least nine volcanoes erupted as *Voyager* flew by the moon. *Galileo*'s flyby of Ganymede captured images of a fractured icy crust. Ganymede may have tectonic plates but they do not drift and collide across the face of the frozen moon.

Some astronomers believe that Ganymede, one of Jupiter's many moons, may have active tectonic plates.

Chapter 4

Plate Tectonics and People

Each year tectonic events such as earthquakes, tsunamis, and volcanic eruptions kill or injure thousands of people around the world. Even more are left homeless. Costs can run into the billions of dollars to repair and rebuild buildings, bridges, roads, pipes, and other infrastructures. Yet, as destructive as these natural disasters are, plate tectonics is also beneficial to humans and living things.

Plate tectonics created climate systems. Atmospheric gases escaped from deep within the earth through surface volcanic eruptions and deep ocean vents. As Earth's atmosphere stabilized over billions of years, the water, nitrogen, and oxygen cycles began. Thus tectonic activities helped to create the necessary conditions for life on the planet and continue to affect life today: Earthquakes, tsunamis, and volcanoes have a direct impact on humans and their societies across the globe.

Earthquakes and People

Along the Ring of Fire, through the Middle East and southeastern Europe, cities and towns, both modern and ancient, have been ravaged by earthquakes. Seismic waves spread out from the epicenter of the quakes and cause buildings to tumble down, bridges to collapse, and railroad tracks to buckle. As more people populate the planet, the chances of damage and death increase. The bustling city of San Francisco along the western coast of California is an example of living in an active tectonic region.

The inhabitants of San Francisco are accustomed to earthquakes. Since its founding the city has been frequently wracked by tremors. One of the worst happened during the early morning hours of April 18, 1906. The city's residents awakened to a terrible sound "like the roar of 10,000 lions."[37] John Farish was a visitor staying at the St. Francis hotel. He remembers that morning:

> I was awakened by a loud rumbling noise which might be compared to the mixed sounds of a strong wind rushing through a forest and the breaking of waves against a cliff. In less time than it takes to tell, a concussion, similar to that caused by the nearby explosion of a huge blast, shook the building to its foundations and it began a series of the most lively motions imaginable. Accompanied by a creaking, grinding, rasping sound, it was followed by tremendous crashes as the cornices of adjoining buildings and chimneys tottered to the ground.[38]

A terrible fire swept through the city following the quake. The Ham and Egg Fire as it came to be known added to the damage caused by the quake. In all, four square miles were leveled. An estimated three thousand people died in either the quake or the fire. Nearly half of the city's population, or 225,000 people, were left homeless.

Seismologists have known since 1906 that quakes would return to San Francisco. The rocks along the San Andreas Fault would someday slip again. Though relatively strong earthquakes shook the city over the next eighty years, the Loma Prieta quake of 1989 brought back the memories of 1906 disasters.

On October 17, 1989, a magnitude 6.9 earthquake ravaged the San Francisco and Monterey Bay areas. Sixty-two people died and 3,757 were injured, and the area sustained $6 billion in damage to buildings, homes, roads, and bridges. The worst damage happened when multileveled overpasses "pancaked," or collapsed, on cars and trucks trapped on the ruined highways. While many buildings were damaged in the quake, those that had been built with newer quakeproof materials and methods survived. Not all modern cities, however, are as well prepared as San Francisco to face an earthquake. Tangshan, China, suffered one of the twentieth century's most devastating earthquakes.

Buildings in downtown San Francisco lie in ruins after a magnitude 6.9 quake struck the Bay Area in 1906.

Tangshan, the Deadliest Earthquake of the Twentieth Century

At 3:42 A.M. on July 28, 1976, a magnitude 7.8 earthquake struck the sleeping city of Tangshan, China, northeast of Beijing. Aftershocks of magnitude 7.1 followed. Over a million people lived in Tangshan near the coast of the Bo Hai, an arm of the Yellow Sea. The earth shook for fourteen to sixteen seconds. When it was over, 93 percent of the city's homes and 78 percent of its industrial structures were completely destroyed. Transportation into the area was impossible. Bridges had collapsed, railroad lines had twisted, and deep cracks cut across roads. It was the human loss, however, that earned the quake its place in history: 242,219 people, or 12.8 percent of the city's inhabitants, were killed.

The Tangshan quake caught seismologists by surprise; its epicenter was not along a fault or plate edge. If scientists had been able to forecast an impending earthquake in the area in a timely fashion, lives and property might have been saved. Therefore, scientists continue to refine their methods and technology in order to predict the next earthquake.

Forecasting Earthquakes Using Gaps and Creeps

Geologists and seismologists use measurements from a variety of sensors such as seismographs, creep meters, and stress meters as well as information from satellites to help them look for patterns of seismic activity. The history of a fault zone can also point to an impending quake. By combining historical and current information scientists can sometimes forecast future earthquakes. Seismic gaps, or periods of inactivity, may be one signal of an impending quake.

Blocks of crust along faults can either move in a slow, steady motion called creep, or they can jolt in

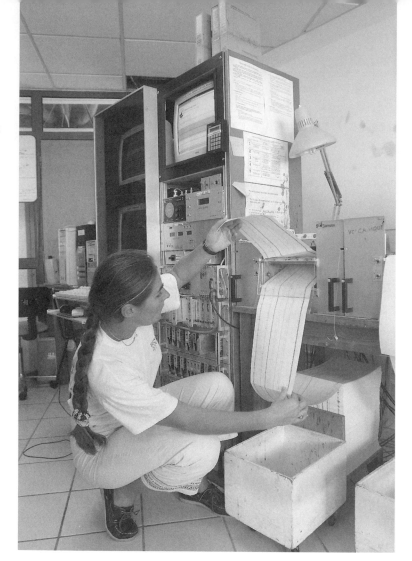

A seismologist on the island of Guadeloupe analyzes seismograph readings to help predict seismic activity in the area.

a sudden, violent movement. Creeping movements cause harmless small to moderate earthquakes. At times the creeping areas along a fault become stuck. Seismic activity drops, and a seismic gap results. Enormous pressure can build up along the locked segments as they stay in position for tens or hundreds of years. Eventually the blocks will snap and jump past each other. When this happens the energy stored in the rocks along the fault is abruptly released in catastrophic earthquakes.

Scientists use human records such as newspaper articles and diaries as well as tree ring dating to piece together the historical timeline of seismic activity along a fault segment. By plotting a segment's peri-

ods of creeps and gaps, seismologists hope to predict when an earthquake will most likely occur.
Even though the art and science of predicting earthquakes has improved over the last three decades,
quake-resistant building methods and materials help
to prevent the collapse of buildings and loss of life.

Construction Techniques and Earthquakes

The quality of construction in an earthquake zone
can mean the difference between life and death. For
example, when a magnitude 6.5 earthquake struck
San Simeon, California, between San Francisco and
Los Angeles in 2003, two people were killed and
forty buildings collapsed. Yet just four days later,
when a quake with the same magnitude struck Bam,
Iran, thousands died and most of the town was destroyed. According to Farzad Naiem, an earthquake
engineer, the quality of construction explains the
extreme difference in loss of life and property between the two regions:

> The construction quality in Iran, particularly in
> rural areas and villages, [is] poor. They have
> been poor for many, many years, and they con-
> tinue to be poor. To give a point of comparison,
> in California, which basically has the same type
> of seismicity as we have in Iran, over the 20th
> century a total number of about 1,600 people
> have died because of earthquakes. In Iran, that
> number over the 20th century is more than
> 126,000, and I don't think anything else but the
> quality of the construction is the reason for the
> difference.[39]

In California most buildings have either wood or
steel and reinforced concrete frames while Iranian
buildings have walls, ceilings, and even roofs made
of heavy and poorly reinforced stone and brick. The
mud brick homes of Bam collapsed and crushed

their sleeping owners. Graham Payne, who participated in a British search and rescue effort on-site at Bam, describes the effect of a quake on mud brick buildings: "As the buildings have collapsed, the mud has gone down to sand. This is what killed so many people."[40] Better construction methods will help cities survive earthquakes. Simple, low-cost materials can make a huge difference in poorer nations.

Quakeproof Buildings

Simple changes in building practices can improve the chances for survival in quake-prone areas. For example, floors on the second level of two-storied buildings are often attached to walls with only nails. By adding inexpensive steel straps that anchor the floor to the walls, the building is less likely to collapse. Also, in many quake zones buildings are topped with heavy tiles, concrete, or mud roofs. These roofs collapse during a quake and crush everyone and everything in the building. By replacing heavy tiles, concrete, and mud with lightweight materials such as corrugated metal roofs, lives can be saved.

Adobe, or mud brick, is a cheap and popular building material in many parts of the world; yet the adobe walls of houses, mosques, and schools came crashing down in Bam when the 2003 quake struck. Adobe construction techniques can be easily modified to prevent such an enormous loss of life. For example, Nader Khalili, an Iranian-born architect, has combined the traditional adobe method with another common material, barbed wire. Khalili lays strands of barbed wire between layers of sandbags filled with wet mud. The barbed wire presses into the sandbags and locks the layers together. Walls built with Khalili's "adobe Velcro" do not crumble, and according to the architect are "as strong as anything one can find anywhere."[41] Nevertheless, in much of the world cities must make do with frail, aging infra-

structures, and especially in impoverished regions an earthquake's tremors can bring down entire city blocks. Moreover, seismic waves can generate another kind of natural disaster—tsunamis.

A truck lies crushed beneath a pile of adobe bricks after the 2003 Bam earthquake. The weak mud bricks used in most of Bam's buildings crumbled in the powerful quake.

Towering Walls of Water

Tsunamis can wreak havoc along every shore of Earth's continents because these great sea waves are created by any sudden disturbance in the ocean. Underwater landslides, volcanic eruptions, explosions, and even the impacts of meteorites can send a chain of waves racing across an ocean or crashing into a nearby shore. Pacific coastlines and islands are especially vulnerable to these giant waves. In fact, 80 percent of the world's tsunamis occur in the Pacific. The islands of Hawaii have often been the target of tsunamis that cross the entire ocean to reach their shores.

In 1946 a magnitude 8.6 earthquake occurred off the Aleutian Islands of Alaska. Nicknamed the April

Fool's Day Earthquake, the quake set off a tsunami. Within minutes a towering wall of water more than one hundred feet tall destroyed a nearby U.S. Coast Guard lighthouse and then raced out into the open waters of the Pacific. At speeds of over five hundred miles per hour, the first of seven waves with heights of twenty-six to thirty-six feet reached Hilo, Hawaii, in less than five hours. The loss of life and property was devastating.

The churning waters of the tsunami drowned 165 people, including 16 school children and 5 of their teachers. On the Laupahoehoe Peninsula twenty-five miles north of Hilo, a school and its teacher cottages overlooked the blue waters of the Pacific. On the morning of April 1, just as the children were arriving, the approaching tsunami sucked the waters from the beach below the school. A group of students scrambled down to investigate but soon were trapped by enormous waves. Other students and

A street in Hilo, Hawaii, is littered with rubble after a devastating tsunami struck the island in 1960. A magnitude 8.6 quake in Chile generated the huge waves.

teachers higher up on the bluff tried to escape. Years later, one student, Masuo Kino, spoke of his experience: "The wave flipped me over and carried me toward the lava rock wall that rimmed the school. I recall telling myself, 'Gee, I'm going to die. I'm going to hit headfirst into that rock wall and I'm going to die.' But miraculously, part of the wave that preceded me smashed into the wall but I was tumbling along, rolling along with all the rocks."[42] Hilo also lost lives and buildings in the waves.

Many people in Hilo were not killed by the initial force of the waves. Instead, they died in the debris-filled waters. Kapua Heuer of Hilo remembers the waves rushing through his town: "I looked out here and saw this great big black wall coming in like this. . . . The noise was terrific, the rolling. . . . And then you heard the screaming. You look and people were stomping trying to reach earth, trying to get out. Dogs swimming around. Then came the crash. . . . Well, it hit buildings, the lighthouse, and the railroad track, and everything. . . . And I said, 'Oh, that's goodbye to Hilo.'"[43] Since the April Fool's disaster, Hawaii has been struck by other tsunamis. After rebuilding from the April Fool's tsunami, Hawaii was once again struck by disaster. On May 22, 1960, a magnitude 8.6 quake off the coast of Chile sent giant waves bearing down on the islands again.

The tsunami's huge waves first ravaged the coast of Chile. The remote fishing villages lacked communication and roads so the death toll could only be estimated. Perhaps three hundred to two thousand people died in waves up to eighty-two feet high. From Chile's devastated coast, the giant waves headed west toward Hawaii. Like enormous ripples, the waves traveled at a speed of two hundred miles per hour and reached Hawaii in less than fifteen hours. The tsunami's thirty-five-foot waves killed sixty-one people. After Hawaii the killer waves continued across the Pacific and reached New Zealand, the Philippine Islands, and Japan. In less than a day the tsunami traveled across the entire Pacific Ocean. While tsunamis have to travel thousands of miles to reach shores, others strike nearby coastlines.

Indian Ocean Catastrophe

On the morning of December 26, 2004, a deadly tsunami swept over the coasts of the Indian Ocean. The tsunami was set off by a magnitude 9.0 earthquake along the boundary between the Indian and Burma plates near the island nation of Indonesia. Off the southwestern coast of Indonesia, the denser Indian plate subducts beneath the much smaller Burma plate near the deep ocean Sunda Trench. Early Sunday morning, the lighter Burma plate suddenly thrust upward, lifting up as much as sixty feet. The jolt displaced trillions of tons of water. In seconds, massive waves pushed out from the quake's epicenter toward heavily populated coastal communities.

Traveling hundreds of miles per hour, the tsunami first crashed into the western coast of Sumatra. This region suffered the greatest number of deaths; nearly 230,000 people died in Indonesia's Aceh and North Sumatra provinces. The waves then rippled across the Indian Ocean. Next hit were the coasts of Thailand, packed with holiday visitors. Across the Indian Ocean lay Sri Lanka, another favorite winter vacation spot, and the eastern coast of India. In less than two hours after the quake, the waves struck those shores. The tsunami continued westerly and washed over the Maldive Islands before reaching the eastern coast of Africa. Though the exact number of dead and missing may never be known, it is believed that nearly 300,000 people perished in the tsunami, and over 1 million people lost their homes.

While the 2004 Asian tsunami killed more people than any other tsunami in recorded history, it was a minor event from a geological standpoint. Simon

Winchester, the author of a book about the explosion of Krakatoa, summarizes the quake's place in tectonic history: "The earth shrugged for a moment. Everything moved a little bit."[44]

Some believe that a tsunami warning system in the Indian Ocean might have prevented the huge number of deaths. While scientists monitored the December 26 earthquake, the terrible waves went unrecorded until they reached shore, and no formal tsunami warning was issued. Had the earthquake occurred in the Pacific Ocean, however, a tsunami warning would have been sent out. Unlike the Indian Ocean, tsunamis are frequent in the Pacific, and Pacific Rim countries have taken steps to prevent or reduce tsunami-related deaths through the Pacific Tsunami Warning System.

Early Warning Systems

The Pacific Tsunami Warning System, or PTWS, is a cooperative effort of twenty-five countries that monitors earthquakes and issues tsunami warnings, watches, and bulletins. Early warning systems like the PTWS rely on a combination of observations from field teams, information from 150 monitoring stations scattered throughout the Pacific, and powerful models created by computers.

Through the digital models of tsunamis, scientists can determine what conditions generate these seismic events, how they travel through the open ocean, and what happens when the waves reach the shallow waters of a coast. Models also help researchers predict a tsunami's level and area of destruction. Armed with this information, geologists, meteorologists, computer experts, and disaster management staff at the PTWS can issue warnings, watches, and bulletins to local response teams that are activated in emergencies. The efforts of the PTWS staff are paralleled by earth scientists who specialize in the prediction of volcanic eruptions.

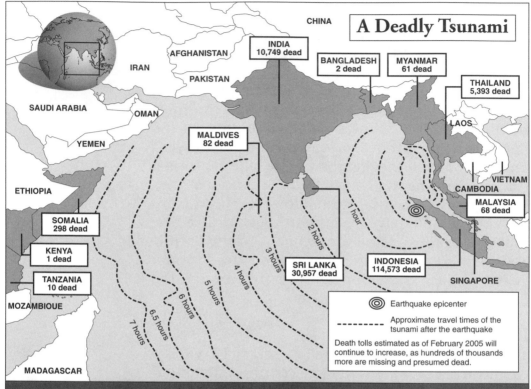

A Deadly Tsunami

On December 26, 2004, the Burma plate (a very small tectonic plate located between the Eurasian and Indian plates) suddenly jolted about 60 feet, causing a 9.0 magnitude earthquake centered off the coast of northern Indonesia. The plate movement also caused the seafloor to be uplifted and shifted by several feet, generating a series of large waves called tsunami. The tsunami traveled thousands of miles across the Indian Ocean, generating waves on the coasts up to 50 feet high, killing hundreds of thousands of people and causing massive destruction.

Growing Need to Predict

Each year the world's population edges ever closer to volcanoes. With 500 million people living within the shadow of a volcano somewhere in the world, some scientists predict the death toll of an unexpected eruption could reach hundreds of thousands. The need for accurate and timely forecasts of volcanic eruptions drives geologists like Stanley Williams to risk their lives hiking down into calderas, or craters, of active volcanoes to take samples and readings. In his book *Surviving Galeras* Williams, a geologist with Arizona State University, explains why these earth scientists are willing to face such dangers: "We all used different methods, but our goal was the same to understand what makes a volcano tick, to forecast eruptions to save lives. We all wanted to save lives."[45]

Scientists can forecast eruptions under certain conditions, which include having a thorough understanding of a volcano's history, an array of sensors in place on the volcano, and the time to analyze information continuously provided from the monitoring stations. Even when these conditions are met vulcanologists can only offer probabilities, not certainties, that an eruption will occur. They cannot predict exactly when a volcano will erupt, nor can they predict the severity of the event.

Monitoring a Volcano

Vulcanologists use a variety of methods and devices to look for signs of volcanic activity. One of the most valuable techniques is the monitoring of a volcano's earthquakes. As magma moves beneath the volcano, recording devices called seismographs plot data giving the magnitude and location of the earthquakes. The more seismographs on a volcano, the more a complete picture emerges of the volcano's interior. Tiltmeters also help scientists see beneath the volcano's surface by detecting tiny changes in the volcano's shape as rising magma makes bulges in the sides of a volcano.

While some research teams continue to descend into a volcano's caldera, the U.S. Geological Survey uses synthetic aperture radar to eliminate the dangers of climbing into an active volcano. This fully automated ground deformation detection system is mounted on a satellite or a plane. As it passes over a volcano the radar sends information about the volcano's ground movements to instruments monitored by research teams. Vulcanologists like Dan Dzurisin with the USGS's Volcano Hazards Program see the automated system as "a tremendous tool because it gives a complete map of ground movements, and we don't have to go into the field to get it."[46] While vulcanologists and their research teams can issue predictions about the probability of an eruption, teams of emergency management professionals are responsible for creating ways to help people survive a volcano's fury.

Disaster Management

A complete emergency management plan can help save lives in a volcanic eruption. An example of a successful program can be found on the Papua New Guinean island of New Britain, where a three-part plan was created during the 1980s. First, scientists and community leaders educated residents about the dangers of a volcanic eruption. Second, hazard maps were posted throughout the island, showing which areas were most likely to be in the path of destruction and which areas were safe. Finally, everyone on the island practiced an emergency evacuation. The plan worked so well that when people noticed signs

Vulcanologists measure a widening crack in the floor of a crater on Mount Saint Helens. Monitoring cracks and bulges can help predict eruptions.

of a possible eruption on September 18, 1994, thirty thousand people chose on their own to evacuate to the safe areas. In the city of Rabaul, 75 percent of the houses collapsed, but only three people died during the eruption.

The recent experience of Rabaul shows how well a potentially disastrous volcanic eruption can be handled, given modern technology, an effective program of public education, and a small, cooperative population. History, however, records many instances of catastrophes due to tectonic activity, and volcanoes are often the cause.

A Calculated Risk

Volcanic rock and ash weather into fertile soil, and many people believe that the risk of living near an active volcano is worth the opportunity to farm some of the world's richest lands. For example, in Indonesia's volcanic soil three crops per year can be grown and harvested. These rice farms are capable of feeding millions of people. The lush vegetation and beautiful scenery also draw settlers to a volcano's slopes. There is, however, a price to pay for living near a volcano.

In 1815 Mount Tambora on the Indonesian island of Sumbawa unexpectedly exploded with a force scientists say was 150 times more powerful than that of Mount Saint Helens. Before the eruption Tambora, with an elevation of 14,100 feet, was the highest mountain in the East Indies. After the eruption it stood at only 9,383 feet above sea level; the eruption blew off 4,700 feet of the mountain's top. Besides the geological and ecological devastation, the eruption also caused the deaths of thousands, and not just in the surrounding region. The Tambora explosion affected people around the world.

The final toll of deaths caused by Tambora is difficult to reach. About twelve thousand people were immediately killed as pyroclastic flows raced down

The enormous volcanic crater of Mount Tambora is clearly visible from space. In 1815 the volcano erupted with explosive force, killing thousands of people.

the mountain and tsunamis sped out from the island. Months after the eruption forty-four thousand to one hundred thousand people died from starvation and disease when their crops and livestock were smothered under thick layers of ash. Even more people may have perished later. The volcano spewed out 200 million tons of gases. The enormous amount of aerosols blocked the sun, and global temperatures dropped by five degrees Fahrenheit. As the planet's climate temporarily cooled, summer frosts killed crops in the northeastern United States and Europe during what came to be known as the "Year Without a Summer."

The catastrophic effects of the eruption of Mount Tambora stick in the mind, and indeed tectonic events like earthquakes, tsunamis, and volcanic eruptions are often harmful to humans and other living things. However, the forces of plate tectonics can also benefit people.

Natural Resources, Earth's Gifts

These forces produce the natural resources essential to civilization, both modern and ancient. Many of the planet's natural resources are created by tectonic forces. Volcanic rocks break down into fertile soils. Far below the surface of Earth immense heat and pressure convert the remains of plants and animals to petroleum and natural gas. Over millions of years coal also forms from plant material that is buried under thick layers of sediments. Plate tectonics is also responsible for the formation of Earth's minerals.

Ancient volcanic activity helped to create some of the world's most valuable minerals. Deposits of copper, gold, silver, lead, and zinc are often found deep within the roots of extinct volcanoes, and diamonds can be found in prehistoric volcanic vents. Ancient ocean crust is also rich in natural resources. In Cyprus, for example, rich copper deposits lie exposed. The oceanic crust was scraped off and piled up as the heavier ocean plate subducted beneath the lighter continental one. Cyprus was such an important source of copper that the Romans called copper by the Latin word *cyprium*, which means "Cyprian metal."

Chapter 5

Plate Tectonics and the Future

Today, the theory of plate tectonics continues to evolve and change. Alfred Wegener anticipated future breakthroughs that would add to or alter this theory and offered these words: "We have to be prepared always for the possibility that each new discovery, no matter which science furnishes it, may modify the conclusions we draw."[47] Though scientists can measure the direction and speed in which the tectonic plates move, they are still constructing an encompassing theory that explains why the plates drift. While scientists know that heat and gravity are the driving forces of plate tectonics, research now focuses on the exact mechanism that moves the continents. Mantle currents, the slab pull of tectonic plates, the ridge push of mid-ocean ridges, and deep mantle superplumes may all play a part in moving Earth's lithospheric plates.

Mantle Currents

Below the crust of Earth, intense heat from the core churns and stirs the molten rock of the mantle like water boiling in a huge cauldron. Hotter, less dense magma rises through the mantle until it bumps up against the colder, solid lithosphere. Nevertheless, some magma may escape through cracks and fissures. If the magma remains trapped below the sur-

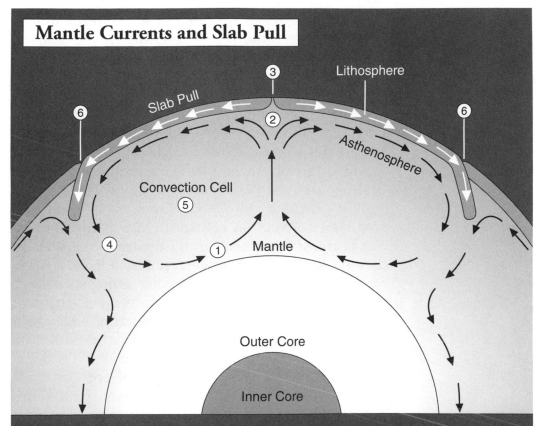

Mantle Currents and Slab Pull

Below Earth's crust, intense heat from the core churns the molten rock of the mantle into circular currents. Geologists once believed these mantle currents helped move the tectonic plates above them. Today it is believed that the plates move on their own, helped by the force of gravity in a process called slab pull.

① Molten rock (magma) is heated by Earth's hot core.

② As the magma becomes hotter, it becomes less dense, so it rises through the mantle until it bumps against the colder, solid lithosphere.

③ Some magma escapes through cracks and fissures in the lithosphere or at ridges or rifts (the spreading edges of divergent tectonic plates).

④ Most magma does not escape but moves along the underside of the tectonic plates until it cools and sinks back into the lower mantle.

⑤ The rising and falling circular currents of magma are called convection cells.

⑥ When an oceanic plate slides underneath another, it is too cold and dense to melt right away. Gravity pulls the heavy plate down into the hot liquid mantle. This process of plate movement is called slab pull.

face, however, it spreads out and sticks to the underside of the tectonic plates. Eventually, the magma cools and begins to sink back into the mantle where it will be warmed and start to rise again. These circular rising and falling currents are called convection cells.

Scientists once believed that huge convection cells acted like giant conveyer belts that carried the tectonic plates along the surface. Lawrence W. Morley, a Canadian geophysicist, saw the mammoth plates as "[riding] around like so much scum on the top of a boiling porridge pot."[48] However, earth scientists now view convection cells primarily as the planet's way of transferring heat from the core to the surface; they no longer consider convection cells as a main force in plate movement. Instead, geologists believe that the plates create their own movements through the pull of gravity on their subducting edges in a process called slab pull.

Slab Pull in the Western Hemisphere

Slab pull occurs when an ancient oceanic plate slides down under its own weight into the asthenosphere and begins to dive deep into the mantle. The subducted plate, however, is too cold and dense to mix with the surrounding rock and sinks slowly into the mantle like a marble in a bowl of thick syrup. In the slab-pull model, the huge blocks of descending crust generate their own force. Gravity pulls both the subducting plate and the overriding plate down into the hot mantle, much as it drags to the floor a heavy cloth laid out too close to the edge of a table.

The speed at which a plate sinks into the mantle is closely tied to the proportion of its margin that is subducting. The subduction zone of the Pacific and Cocos plates involves 40 percent of their edges; consequently, both plates move at a rate of two inches per year. On the other hand, the North America plate has a much smaller subduction margin and moves at a rate of only one-half an inch per year.

Slab pull also accounts for the downward movement of the center of North America. Geological evidence indicates that approximately 100 million years ago the ancient Farallon plate dove beneath the west coast of North America at a very shallow angle and at such a quick rate that the Farallon plate subducted beneath the continental plate faster than it had formed. As the cold oceanic plate pushed under the continental plate, the western region of North America was elevated. However, as the Farallon plate continued to burrow under North America, it finally began to slide into the mantle on its journey to the center of the continent. The sinking oceanic slab created downward currents in the hot, liquid mantle and began to pull at the interior of North America. Over millions of years the center of the continent dropped several thousand feet, and a shallow sea about six hundred miles across flooded parts of present-day central Colorado. Eventually North America broke free of the descending plate and rose again to its present elevation. The Farallon plate now lies one thousand miles below the eastern coast of the United States.

Slab Pull in the Eastern Hemisphere

North America, however, is not the only continent affected by slab pull. Australia also experienced the downward force of a sinking plate. Using computer simulations, geophysicists theorize that an enormous subduction zone surrounding the supercontinent of Gondwanaland had caused Australia to bob up and down. About 130 million years ago Australia broke away from Gondwanaland. As it passed over the edges where cold, oceanic slabs were sinking into the mantle, the eastern section of Australia dropped, and seawater flooded the region. Then, about 70 million years ago as the continent continued on its eastward journey away from the slab, Australia popped back up to its original elevation.

Presently, Australia is sinking once again. This time Indonesia is the culprit. Indonesia lies at the intersection of an enormous, present-day subduction zone in the Pacific and Indian oceans. As the plate that Indonesia rides upon continues to sink, Australia is pulled down along with it.

Another mechanism for plate movement, not seen as important as slab pull, is called ridge push. When hot, buoyant magma rises up through one of the oceanic ridges, it cools at the surface and slides down the sides of the ridge. Newer lava pushes out from the ridge and forces the older cooling rocks to move away from the ridges. The amount of movement related to ridge push depends on the elevation of the ridge; newly formed rocks slide down on a steep slope faster than on a gentle one.

The Mantle Plume Phenomenon

Whereas ridge push and slab pull explain in part how the plates move over the asthenosphere, many scientists are convinced that mantle plumes, especially enormous ones called superplumes, also play an important part in the shaping of Earth's surface.

Mantle plumes are long blobs of magma that are slowly rising from the bottom of the mantle. Shaped like narrow columns, these plumes originate eighteen hundred miles beneath the surface at the boundary of the core and mantle. In this intensely hot region, portions of mantle material become hotter and therefore lighter than surrounding rock. The more buoyant rock begins to rise through the mantle like a hot air balloon.

Most plumes are a few hundred miles across and rise at a rate of approximately six feet per year. Over millions of years a plume develops with a large head followed by a slender tail that reaches back to the core-mantle boundary. When it reaches the lithosphere, the plume's head flattens and spreads out against the rigid layer of cold rock. Hot spots, volca-

Mantle Plumes

Mantle plumes are long narrow blobs of molten rock (magma) that slowly rise from eighteen hundred miles below Earth's surface, at the superhot boundary of the core and mantle. When it reaches the cool lithosphere, the plume's head flattens and spreads out, sometimes breaking through the crust through fissures and cracks in the lithosphere. These breakthroughs can result in volcanoes or hot spots at Earth's surface.

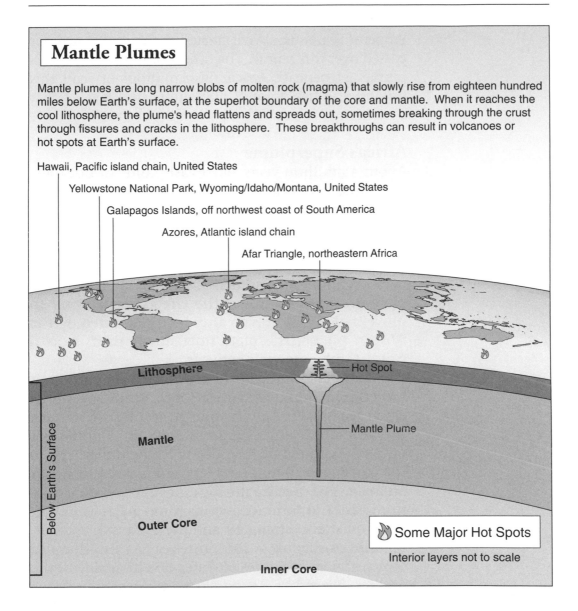

Hawaii, Pacific island chain, United States

Yellowstone National Park, Wyoming/Idaho/Montana, United States

Galapagos Islands, off northwest coast of South America

Azores, Atlantic island chain

Afar Triangle, northeastern Africa

Lithosphere

Hot Spot

Mantle Plume

Below Earth's Surface

Mantle

Outer Core

Some Major Hot Spots

Interior layers not to scale

Inner Core

noes, and flood basalts, which are vast sheets of lava, can result when magma from the head forces its way to the surface.

While scientists speculate that there are about twenty deep mantle plumes, recent three-dimensional images show two huge superplumes of rock are rising on opposite sides of the planet. One superplume is located beneath the Pacific Ocean. The Hawaiian Islands and

Emperor Seamounts were created as the Pacific seafloor passed over this plume. The other continent-size superplume lies beneath Africa. Over millions of years this superplume has dramatically altered the face of that continent.

Africa's Superplume

About 45 million years ago a vast plume of hot rock, 750 miles across and 900 miles high, rose through the mantle to rest beneath the African plate. When the superplume bumped into the roots of the plate, the continent was snagged on the huge mushroom-shaped plume. Africa has never been a fast-moving plate. Without any bordering subduction zones, the plate did not experience the effects of slab pull to move it along. However, after being captured by the African plume, the plate now slowly drifts northward at only one inch per year.

As the superplume pushes its way to the surface, it has fashioned some of Africa's most striking geological features. The jumbled topography of eastern Africa resulted when a plume, possibly an offshoot of the African plume, collided with the underbelly of the continent 30 million years ago beneath the Afar Triangle. The area fractured, and Arabia broke away from Africa to form its own tectonic plate. The shattered crust also stretched and dropped in elevation. Seawater rushed in to form the Red Sea and the Gulf of Aden, narrow bodies of water on either side of the Arabian Peninsula. As the plume continued to shove its way up, a jagged tear formed the East African rift. The plume is still traveling upward and splitting the continent. Two new separate plates are developing on the eastern edges of the rift system, the Nubian plate to the west and the Somalian plate to the east.

The superplume under Africa has also pushed up the southern tip of the continent to create an enormous plateau, more than one thousand miles across and almost a mile high. Intense heat from the ascending plume also melted the bottom of the African plate, and magma pushed its way to the surface and erupted in a flood of basalt. Though rare, flood basalts occur on all continents, and mantle plumes explain how they formed.

Africa's unique landscape has been shaped by its superplume. To study the puzzling topography of Africa, scientists use seismic surveys. In January 2003 seventy scientists from Europe, the United States, and Ethiopia participated in the world's largest three-dimensional seismic survey. They detonated nineteen explosions throughout Ethiopia's Rift Valley as a part of Project EAGLE, the Ethiopian-Afar Geoscientific Lithospheric Experiment. New images of the African superplume, combined with EAGLE's data, show a continent that is not simply cracking up. Instead, the project confirmed that Africa is slowly stretching and thinning along the eastern rift valleys.

The Yellowstone Plume

Other plumes lie under Earth's crust. Though not as large as the African and Pacific plumes, these smaller ones also mold the surface terrain. A slow-moving mantle plume exists beneath the North American plate. Dubbed the Yellowstone plume, it has created a spectacular array of geological features scattered throughout the western region of the United States.

The Yellowstone plume activated an ancient volcanic system that lies in the heart of the northwestern United States. The Columbia River Plateau, part of that system, covers nearly 2 million square miles in Washington and Oregon. The plateau formed when great floods of basalt, some over one mile thick, covered the land. Southeast of the Columbia Plateau, the Snake River Plain slices through the mountains of Idaho. It too is covered with basalt lava flows as well as small shield volcanoes. To the northwest, the Yellowstone Plateau rises to an elevation of eight thousand feet above sea level. In the

Yellowstone National Park is situated atop a mantle plume. The plume generates steamy hot-water geysers and bubbling mudpots.

middle of Yellowstone National Park, three huge calderas testify to the massive volcanic eruptions that occurred only 620,000 years ago.

Even today the plume beneath Yellowstone produces dramatic though less ferocious results. As the Yellowstone plume heats up the surrounding rock, groundwater trapped in the rock layers also becomes hot and rises to the surface. The superheated water feeds Yellowstone's spectacular geysers, its boiling hot springs, and bubbling mudpots. In fact, when Old Faithful, Yellowstone's most famous geyser, sends its spray of scorching water 180 feet in the air, tourists are witnessing firsthand the effects of tectonic forces.

Three Competing Tectonic Theories

Plate tectonics alone does not explain the plateau of southern Africa or the world's flood basalts. However, "plume tectonics," as some scientists refer to the effects of rising superplumes, does help to fill in some of the gaps of the older theory. In the future, new technologies will add to the overall picture of Earth, and plume tectonics might combine with plate tectonics to show how plates move both horizontally and vertically. Some scientists, though, are still doubtful that the plumes are responsible for the rise of continents. Bradford Hager of the Massachusetts Institute of Technology states, "I thought it was nuts at first." But after reviewing his own models, he believes it "is probably a neat idea but I wouldn't bet the farm on it yet."[49]

While plate tectonics has revolutionized and unified the field of geology, it still has its flaws. The theory clearly describes and explains tectonic activity associated with plate boundaries and ocean geology but it does not answer many questions about the continents and geological activity in their interiors. Continents do not always follow the rules of plate tectonics. Geologists want to know why and how the interior of continents changes shape.

Today, earth scientists have moved from studying the edges of the plates to exploring the mechanics of continental tectonics. One place under study is the Tibetan Plateau, a raised area that was formed when India slammed into Asia. Instead of behaving like a rigid plate, recent GPS measurements and images from an advanced form of synthetic aperture radar show Tibet to be bending.

Ancient Mysteries

The Tibetan Plateau presents a mystery that plate tectonics does not seem to be able to solve. Over the last 40 million years the impact when India slammed into Asia compressed many square miles of land in Asia into giant furrows and forced up the Himalayan Mountains and the Tibetan Plateau. However, the oldest part of the Indian plate survived while sections of Asia buckled and crumpled under the force of the crash. So, while ocean plates appear to be strong and unbending, continental plates like Asia have demonstrated the capacity to wrinkle and warp. Also, the two colliding continents squeezed the plateau of Tibet much like a slab of soft clay between two blocks of wood. Today, the crust of Tibet is forty to fifty miles thick, almost twice the thickness of crust found elsewhere on the planet. This compression of the interior of continents appears to contradict the theory of plate tectonics. The thickness of continental plates also seems to challenge the theory.

Cratons, Earth's oldest land formations, pose additional problems for plate tectonics. These continental cores are billions of years old; dozens formed at about the same time, over 4 billion years ago. North America's seven cratons, found in central Canada and the north central United States, merged about 2 billion years ago, making it the world's oldest continent.

The theory of plate tectonics proposes that continental plates are no thicker than about 60 miles, and that the lithosphere and asthenosphere under both the continental and ocean crusts are basically the same. However, cratons are much thicker than 60 miles and some scientists believe their roots may extend hundreds of miles into the asthenosphere. For decades, the depth of the roots has raised a debate among the earth science community. Recently, however, those measurements are being questioned, and a revised estimate suggests that continental roots may reach down to 125 to 160 miles.

Though cratons are strong, stable shields of rock in the middle of continents, the continental interiors are not always tectonically quiet regions like the rules of plate tectonics imply. The middle of the United States demonstrates that the center of a continent can be a very active place when intraplate earthquakes or quakes, located sometimes hundreds of miles from the closest plate edge, strike the heartland of the United States.

Failed Rifts

In the winter of 1811 a series of catastrophic earthquakes hit the area of New Madrid, Missouri, in the center of the North American plate and far from any plate boundary. The quakes were felt as far away as northern Alabama and Georgia, and thousands of aftershocks followed throughout the spring of 1812. The area was devastated; large sections of land sank and filled with water to form new lakes, the Mississippi River changed course, and 150,000 acres of forests were destroyed. Even today, this region is tectonically active. It has more earthquakes than any other part of the United States east of the Rocky Mountains.

The quakes were the result of movement along an ancient rift called the Reelfoot Rift, part of the New Madrid Seismic Zone that stretches from Missouri and

Astronauts on the International Space Station took this photo of Mount Everest in the Himalayan Mountains, formed when the Indian plate slammed into Asia 40 million years ago.

Illinois into northeastern Arkansas. About 500 million years ago the North American continent began to stretch and split apart, opening up a tear in Earth's crust. However, the rifting process stopped, and the Reelfoot Rift became what geologists call a failed rift. If it had continued, the North American plate would have broken apart and an ocean would have filled in to separate the two newly formed continents. While the rift failed to transform into a mid-ocean ridge, the crust around the rift was fractured into many faults. This loose system of faults and rifts is now buried miles below the surface under thick layers of sediments. Yet occasionally the rocks along the fault zones slip and jump past each other to cause the many earthquakes of the New Madrid Seismic Zone.

Not all failed rift systems continue to be seismically active. The Midcontinent Rift System of North America runs from northern Kansas across the southeastern region of Nebraska, through Iowa, Minnesota, and Wisconsin to the southern end of Lake Superior and is 180 miles wide and over 800 miles long. The rift marks the site where the North American continent began to rip apart about 1 billion years ago. Scientists speculate the rift may have been caused by a rising mantle plume. However, as with the Reelfoot Rift, the tearing process stopped. In both instances, scientists are unsure why the rifts did not continue to spread. Unlike the Reelfoot Rift, however, the land above the Midcontinent Rift is not plagued by earthquakes and tremors. Instead, flat fields of grain and gently rolling hills now cover the enormous crack.

While scientists recreate the geologic past with evidence found in present-day rocks, they often turn to computer modeling in order to predict the future movement of continents. Using time-lapse computer images, the journey of the plates across Earth's surface can be rewound or fast-forwarded like a movie.

The Future Earth

Precise measurements of current direction and speed help geologists to predict the movement of the plates far into the future. While GPS networks, lasers, and computer modeling can aid scientists, predicting the future is not an easy task. Christopher Scotese, a professor of paleogeography at the University of Texas at Arlington, is director of the PALEOMAP Project, which has developed maps of what the world looked like long ago as well as what it might look like in the future. He admits that creating maps of millions of years into the future is difficult. "It's like if you're traveling on the highway, you can predict where you're going to be in an hour, but if there's an accident or you have to exit, you're going to change direction. And we have to try to understand what causes those changes. That's where we have to make some guesses about the far future—150 to 250 million years from now."[50]

Based on current rates and directions, computer models show that in 250 million years today's continents will once again collide, forming into a single super continent. But before this happens, other changes will occur. In 50 million years Earth will be a much different place. Baja California will grind northward, up the western coast of North America, to lodge near the Aleutian Islands. The Scandinavian Peninsula will touch mainland Europe while the boot of Italy will disappear into Greece. Eventually, the Mediterranean Sea will cease to empty into the Atlantic Ocean, as Africa moves northward, squeezing shut the Strait of Gibraltar. After tens of millions of years, most of Africa will be north of the equator, and the Somali plate on the eastern side of the continent will tear away. India will continue to push into the underbelly of Eurasia while Australia will collide with Southeast Asia.

Predictions of the location of the plates become trickier when looking beyond 100 million years into

Future Continental Drift

Based on current directions and rates of speed, geologists predict that several million years from now, the earth's tectonic plates will bring the continents together again.

225 million years ago

65 million years ago

Today

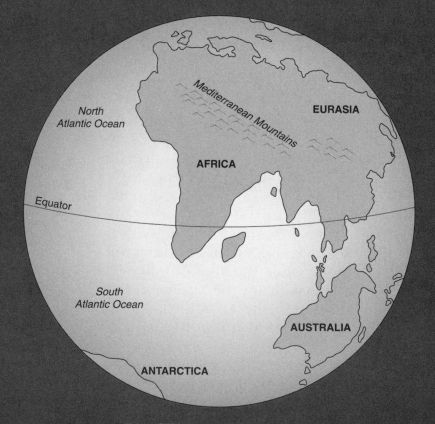

50 million years from today

In this possible future model, Africa will move northward toward the European continent, closing up the Mediterranean Sea, and Australia will merge with the islands of Southeast Asia.

the future. Some simulations predict that in 100 million years, the Atlantic Ocean will begin to close as new subduction zones along the eastern coasts of North and South America begin to consume the sea floor that separates the Western and Eastern Hemispheres. During this time the Caribbean Sea will begin to close. India will plow through Eurasia on its way to the northern edges of the continent. In 150 million years North and South America will be pulled into the current joint landmasses of Eurasia and Africa. In 250 million years a new Pangaea might be born as the continents once again fuse together into a single supercontinent before the inexorable tectonic forces split it into new continents as the cycle of creation and destruction continues.

Evolution of the Theory

With the theory of plate tectonics scientists can peer deep into the past to explain the present and predict the future. Tjeerd H. Van Andel, now an honorary professor of Earth History at Cambridge University in England, believes "Earth science is history, because all that happens today is the product of what happened yesterday. Every feature of the earth is part of a chain of events that reaches deep into the past, each dependent on what went before, each influencing what came after."[51]

New breakthroughs will continue to build a deeper understanding of the world. Canadian geophysicist Lawrence W. Morley is optimistic about the future of earth sciences. In 1963 Morley was one of the first scientists to link the magnetic zebra stripes on the ocean floor with seafloor spreading and drifting continents. Undaunted by the vast amount of new information and studies, he writes forty years later in *Plate Tectonics: An Insider's History of the Modern Theory of the Earth*, "Science has only scratched the surface of the natural world. Opportunities for new important discoveries are limitless."[52]

Notes

Introduction: Modern Tragedies and Ancient Forces

1. Quoted in Karen Wright, "How Earth Rocks," *Discover*, February 2001, p. 22.
2. Quoted in Ali Akbar Dareini, "Thousands Killed in Quake in Bam, Iran," *AP Online*, 2003. www.ap.org.
3. Ellen Prager, *Furious Earth*. New York: McGraw-Hill, 2000, p. 8.
4. Quoted in *Earth Bulletin*, "Quakes from Space," American Museum of Natural History, 2004. http://earthbulletin. amnh.org/A/l/l/index.html.

Chapter One: Continental Drift and Plate Tectonics

5. Eric H. Christiansen and W. Kenneth Hamblin, *Earth's Dynamic Systems*. Upper Saddle River, NJ: Prentice-Hall, 2001, p. 461.
6. Tanya Atwater, "When the Plate Tectonic Revolution Met Western North America," in *Plate Tectonics: An Insider's History of the Modern Theory of the Earth*, ed. Naomi Oreskes. Boulder, CO: Westview, 2003, p. 250.
7. Quoted in David M. Lawrence, *Upheaval from the Abyss*. Piscataway, NJ: Rutgers University Press, 2002, p. 34.
8. Seiya Uyeda, *The New View of the Earth*, trans. Masako Ohnuki. San Francisco: W.H. Freeman, 1971, p. 6.
9. Alfred Wegener, *The Origin of Continents and Oceans*, trans. John Biram. New York: Dover, 1966, p. 77.
10. Quoted in Tjeerd H. Van Andel, *New Views on an Old Planet: A History of Global Change*. New York: Cambridge University Press, 1994, p. 173.
11. Quoted in Lawrence, *Upheaval from the Abyss*, p. 18.
12. Wegener, *The Origin of Continents and Oceans*, p. viii.
13. Quoted in Lawrence, *Upheaval from the Abyss*, p. 165.
14. Quoted in Lawrence, *Upheaval from the Abyss*, p. 157.

15. Atwater, in Oreskes, *Plate Tectonics*, pp. 245–46.
16. Prager, *Furious Earth*, p. 8.
17. Lawrence, *Upheaval from the Abyss*, p. 221.
18. Prager, *Furious Earth*, p. 12.

Chapter Two: The Unified Theory of Plate Tectonics

19. Andrew Alden, "Sounding the Deep Earth," About, 2004. http://geology.about.com/library/weekly/aa030198.htm.
20. Allesandro Forte, "A Four-Piston Engine Drives Earth from the Inside, New Study Shows," University of Toronto, September 2003. http://news.utoronto.ca/bin2/010425a.asp.
21. Prager, *Furious Earth*, pp. 100, 101.
22. Quoted in *Seattle Post-Intelligencer*, "Mount St. Helens: From the Ashes," 2002. http://seattle pi.nwsource.com/mountsthelens/mem2.shtml.
23. Quoted in *Earth Bulletin*, "Quakes from Space."
24. Quoted in *Earth Bulletin*, "Quakes from Space."

Chapter Three: Plate Tectonics Today

25. Quoted in *Earth Bulletin*, "Quakes from Space."
26. Quoted in *Earth Bulletin*, "Quakes from Space."
27. Quoted in Daniel Wood, "Cracking the Crust," *Canadian Geographic*, January/February 1996, p. 51.
28. Quoted in Wood, "Cracking the Crust."
29. Quoted in Beth Daley, "Scientists Bid Farewell to Little Sub That Could," *Boston Globe*, August 7, 2004.
30. Quoted in Chad Cohen, "U.S. Deep-Sea Expedition Probes Earth's Final Frontier," *National Geographic Today*, February 21, 2003. http://news.national geographic.com/news.
31. Cindy Lee Van Dover, "Depths of Ignorance," *Discover*, September 1993, p. 37.
32. Van Dover, "Depths of Ignorance."
33. Quoted in Bijal P. Trivedi, "Using Unmanned Subs to Probe the Deep," *National Geographic Today*, August 16, 2001. http://news.national geographic.com/news.

34. Alan D. Chave and John R. Delaney, "NEPTUNE: A Fiber-Optic 'Telescope' to Inner Space," *Oceanus*, January 2000, p. 10.
35. Chave and Delaney, "NEPTUNE."
36. Quoted in Andrea Dorfman and Michael D. Lemonick, "The Last Frontier," *Time*, August 14, 1995, p. 52.

Chapter Four: Plate Tectonics and People

37. Ellen Klages, "Life Along the Faultline," Exploratorium, 1999. www.exploratorium.edu/faultline/1906/index.html.
38. Quoted in Klages, "Life Along the Faultline."
39. Quoted in *The Newshour with Jim Lehrer*, "Deadly Tremor in Iran," Public Broadcasting Service, 2003. www.pbs.org.
40. Quoted in Sam Lubell and Peter Reina, "Along with Lives, Architectural Treasures Lost in Bam, Iran," *Architectural Record*, February 2004, p. 26.
41. Quoted in Charles W. Petit, "Aftershock," *U.S. News & World Report*, January 12, 2004, p. 42.
42. Quoted in Center for Oral History, "Tsunamis Remembered," University of Hawaii, 2004. www.oralhistory.hawaii.edu.
43. Quoted in Center for Oral History, "Tsunamis Remembered."
44. Quoted in Evan Thomas and George Wehrfritz, "Tide of Grief," *Newsweek*, January 10, 2005, p. 32.
45. Stanley Williams and Fen Montaigne, *Surviving Galeras*. New York: Houghton Mifflin, 2001, p. 10.
46. Quoted in Peter Tyson, "Can We Predict Eruptions?" *Nova*, Public Broadcasting Service, 1998. www.pbs.org/wgbh/nova.

Chapter Five: Plate Tectonics and the Future

47. Wegener, *The Origin of Continents and Oceans*, p. viii.
48. Lawrence W. Morley, "The Zebra Pattern," in Oreskes, *Plate Tectonics*, p. 79.

49. Quoted in Richard A. Kerr, "The Great African Plume Emerges as a Tectonic Player," *Science*, July 1999, p. 187.
50. Quoted in Patrick L. Barry, "Continents in Collision," *FirstScience*, 2004. www.firstscience. com.
51. Van Andel, *New Views on an Old Planet*, p. xi.
52. Morley, in Oreskes, *Plate Tectonics*, p. 85.

For Further Reading

Books

J.S. Kidd and Renee A. Kidd, *On Shifting Ground: The Story of Continental Drift*. New York: Facts On File, 1997. This book traces the development of the theory of continental drift through the stories of the scientists who dared to propose that continents move.

Helen Roney Sattler, *Our Patchwork Planet: The Story of Plate Tectonics*. New York: Lothrop, Lee, and Shepard, 1995. Clear and colorful illustrations and photographs add to the author's lively explanation of plate tectonics.

Carole G. Vogel, *Shifting Shores: The Restless Sea*. New York: Franklin Watts, 2003. Some say the plate tectonic revolution began with the study of the ocean floor. Beginning with the ocean, the author describes how tectonic forces and events have shaped the planet's surface and its life.

Web Sites

Canada's National LITHOPROBE Geoscience Project (www.lithoprobe.ca). The Web site describes how the *LITHOPROBE* project set out to make a geological map of the land beneath Canada.

The NEPTUNE Project (www.neptune.washington.edu/index.html). Nearly one thousand miles of fiber optics and power cables will wire the entire Juan de Fuca plate off the coast of Oregon and Washington. Cameras, underwater robots, and sensors will send back real-time information to classrooms and laboratories.

PALEOMAP Project (www.scotese.com/Default.htm). Animated maps show how Earth has changed over the past 1,100 million years.

United States Geological Survey (www.usgs.gov). This Web site is one of the most comprehensive sources for information about Earth, including its natural and living resources, natural hazards, and the environment.

Works Consulted

Books

Eric H. Christiansen and W. Kenneth Hamblin, *Earth's Dynamic Systems*. Upper Saddle River, NJ: Prentice-Hall, 2001. Using the hydrological systems and the tectonic system, Christiansen and Hamblin's college text demonstrates how much of the physical world relates to these two themes.

David M. Lawrence, *Upheaval from the Abyss*. Piscataway, NJ: Rutgers University Press, 2002. The key to unlocking the mystery of plate tectonics lies at the bottom of the ocean. Lawrence traces the development of the theory of plate tectonics beginning with Wegener and his idea of drifting continents and ending in the 1960s with the discoveries of seafloor spreading.

Naomi Oreskes, ed., *Plate Tectonics: An Insider's History of the Modern Theory of the Earth*. Boulder, CO: Westview, 2003. This book is a collection of firsthand accounts by the men and women who were instrumental in developing the theory of plate tectonics. In their own words, these geologists, geophysicists, oceanographers, and explorers recount their feelings and memories about the times and experiences that led to a revolution in how sciences viewed the world.

Ellen Prager, *Furious Earth*. New York: McGraw-Hill, 2000. In clear and lively language, Ellen Prager explains how and why volcanoes, earthquakes, and tsunamis occur.

Seiya Uyeda, *The New View of the Earth*. Trans. Masako Ohnuki. San Francisco: W.H. Freeman, 1971. Uyeda, a geophysicist, shows how the acceptance of continental drift changed how scientists viewed the world.

Tjeerd H. Van Andel, *New Views on an Old Planet: A History of Global Change*. New York: Cambridge University Press, 1994. In clear and precise language, Van Andel's often humorous narration tells the story of the evolution of Earth and looks into the future as the world faces the challenges of global change.

Alfred Wegener, *The Origin of Continents and Oceans*. Trans. John Biram. New York: Dover, 1966. Written nearly one hundred years ago, Alfred Wegener's original words have now been translated into English. Though Wegener was writing for the earth scientists of his time, many of the concepts presented in this book can be found in the modern theory of plate tectonics.

Stanley Williams and Fen Montaigne, *Surviving Galeras*. New York: Houghton Mifflin, 2001. Stanley Williams barely survived a minor volcanic eruption of Galeras, a Colombian volcano. Compared to other eruptions, this was just a small "hiccup" but for Williams and the other members of his team it was a deadly experience. This autobiographical account tells of his harrowing experience on the volcano and his long road to recovery. Along the way, Williams, a university professor, manages to pack in an enormous amount of information about volcanoes and the life of a geologist.

Periodicals

Alan D. Chave and John R. Delaney, "NEPTUNE: A Fiber-Optic 'Telescope' to Inner Space," *Oceanus*, January 2000.

Beth Daley, "Scientists Bid Farewell to Little Sub That Could," *Boston Globe*, August 7, 2004.

Andrea Dorfman and Michael D. Lemonick, "The Last Frontier," *Time*, August 14, 1995.

Richard A. Kerr, "The Great African Plume Emerges as a Tectonic Player," *Science*, July 1999.

Sam Lubell and Peter Reina, "Along with Lives, Architectural Treasures Lost in Bam, Iran," *Architectural Record*, February 2004.

Charles W. Petit, "Aftershock," *U.S. News & World Report*, January 12, 2004.

Evan Thomas and George Wehrfritz, "Tide of Grief," *Newsweek*, January 10, 2005.

Cindy Lee Van Dover, "Depths of Ignorance," *Discover*, September 1993.

Karen Wright, "How Earth Rocks," *Discover*, February 2001.

Daniel Wood, "Cracking the Crust," *Canadian Geographic*, January/February 1996.

Internet Sources

Andrew Alden, "Sounding the Deep Earth," About, 2004. http://geology.about.com./library/weekly/aa030198.htm.

Patrick L. Barry, "Continents in Collision," *FirstScience*, 2004. www.firstscience.com.

Center for Oral History, "Tsunamis Remembered," University of Hawaii, 2004. www.oralhistory.hawaii.edu.

Chad Cohen, "U.S. Deep-Sea Expedition Probes Earth's Final Frontier," *National Geographic Today*, February 21, 2003. http://news.nationalgeographic.com/news.

Ali Akbar Dareini, "Thousands Killed in Quake in Bam, Iran," *AP Online*, 2003. www.ap.org.

Earth Bulletin, "Quakes from Space," American Museum of Natural History, 2004. http://earthbulletin.amnh.org/A/1/1/index.html.

Allesandro Forte, "A Four-Piston Engine Drives Earth from the Inside, New Study Shows," University of Toronto, September 2003. http://news.utoronto.ca/bin2/010425a.asp.

Ellen Klages, "Life Along the Faultline," Exploratorium, 1999. www.exploratorium.edu/faultline/1906/index.html.

The Newshour with Jim Lehrer, "Deadly Tremor in Iran," Public Broadcasting Service, 2003. www.pbs.org.

Seattle Post-Intelligencer, "Mount St. Helens: From the Ashes," 2002. http://seattlepi.nwsource.com/mountsthelens/mem2.shtml.

Peter Tyson, "Can We Predict Eruptions?" *Nova*, Public Broadcasting Service, 1998. www.pbs.org/wgbh/nova.

Bijal P. Trivedi, "Using Unmanned Subs to Probe the Deep," *National Geographic Today*, August 16, 2001. http://news.nationalgeographic.com/news.

Index

Picture Credits

About the Author

Kay Jackson writes children's nonfiction books about nature, geology, world religions, and geography. She lives and teaches in Tulsa, Oklahoma.